The Cytoskel

The Practical Approach Series

SERIES EDITORS

D. RICKWOOD
*Department of Biology, University of Essex
Wivenhoe Park, Colchester, Essex CO4 3SQ, UK*

B. D. HAMES
*Department of Biochemistry and Molecular Biology, University of Leeds
Leeds LS2 9JT, UK*

Affinity Chromatography
Anaerobic Microbiology
Animal Cell Culture
Animal Virus Pathogenesis
Antibodies I and II
Biochemical Toxicology
Biological Membranes
Biosensors
Carbohydrate Analysis
Cell Growth and Division
Cellular Calcium
Cellular Neurobiology
Centrifugation (2nd edition)
Clinical Immunology
Computers in Microbiology
Crystallization of Proteins and
 Nucleic Acids
Cytokines
The Cytoskeleton
Directed Mutagenesis
DNA Cloning I, II, and III
Drosophila
Electron Microscopy in Biology

Electron Microscopy in
 Molecular Biology
Essential Molecular Biology I
 and II
Fermentation
Flow Cytometry
Gel Electrophoresis of Nucleic
 Acids (2nd edition)
Gel Electrophoresis of Proteins
 (2nd edition)
Genome Analysis
HPLC of Small Molecules
HPLC of Macromolecules
Human Cytogenetics
Human Genetic Diseases
Immobilised Cells and
 Enzymes
Iodinated Density Gradient
 Media
Light Microscopy in Biology
Liposomes
Lymphocytes
Lymphokines and Interferons

The Cytoskeleton
A Practical Approach

Edited by
K. L. CARRAWAY

Department of Cell Biology,
University of Miami School of Medicine
Miami, FL, USA

and

C. A. C. CARRAWAY

Department of Biochemistry and Molecular Biology
University of Miami School of Medicine
Miami, FL, USA

IRL PRESS
—at—
OXFORD UNIVERSITY PRESS
Oxford New York Tokyo

Oxford University Press, Walton Street, Oxford OX2 6DP
Oxford New York Toronto
Delhi Bombay Calcutta Madras Karachi
Petaling Jaya Singapore Hong Kong Tokyo
Nairobi Dar es Salaam Cape Town
Melbourne Auckland
and associated companies in
Berlin Ibadan

Oxford is a trade mark of Oxford University Press

Published in the United States
by Oxford University Press, New York

© *Oxford University Press, 1992*

A catalogue record for this book is available from the British Library

Library of Congress Cataloging in Publication Data
The Cytoskeleton: a practical approach/edited by Kermit L. Carraway
and Coralie A. Carothers Carraway.
(Practical approach series)
Includes bibliographical references and index.
1. Cytoskeleton. I. Carraway, Kermit L. II. Carraway, Coralie
A. Carothers. III. Series.
[DNLM: 1. Cytoskeleton. QH 603.C96 C9965]
QH603.C96C97   1992   574.87'34—dc20   91–20903
ISBN 0–19–963257–X (hbk.)
ISBN 0–19–963256–1 (pbk.)

Typeset by Cambrian Typesetters, Frimley, Surrey
Printed in Great Britain by
Information Press Ltd, Eynsham, Oxford

Preface

THE field of cytoskeletal research has passed its early developmental stages and progressed into a highly dynamic state. Early studies on the cytoskeleton focused on morphologial observations of cellular or isolated cytoskeletal elements and on the isolation and characterization of cytoskeletal proteins. Such characterizations continue to reveal the complexity of the composition and structure of cellular cytoskeletons. However, more recent research has turned toward investigations of the dynamics and functions of cytoskeletal structures. These investigations are driven by advances in three technologies: video microscopy, immunology and molecular biology. The first permits observations on the dynamics of individual cytoskeletal structures *in vitro* and *in vivo*, developing an understanding of cytoskeletal dynamics and interactions unavailable by other techniques. The use of antibodies (polyclonal or monoclonal) has clarified our understanding of the cellular localization of cytoskeletal proteins and provided a tool for intervening in the dynamic behaviour of cytoskeletal elements. The use of molecular biology techniques has provided invaluable information concerning the primary structures of cytoskeletal proteins and their possible structure/function relationships. Moreover, these technologies, along with classical genetic techniques, provide avenues for understanding the organization and complex functions of the cytoskeletal structures. However, the understanding of such structures and their roles at the molecular level requires a secure knowledge of the chemistry of the proteins involved and their interactions. That knowledge should begin to emerge from the three-dimensional structures provided by crystallographic methods and molecular modelling. Biochemical studies are essential to understanding the interactions between the cytoskeletal proteins and between cytoskeletal elements and other cellular components, such as membranes. Integration of results from all of these technologies should yield a picture of cellular dynamics at the molecular level and a much greater appreciation of the role of the cytoskeleton in such complex cellular behaviours as motility, transmembrane signalling, and growth regulation.

A major problem in developing this volume was the complexity of the subject. Obviously, no single volume could describe all of the approaches and techniques of value in cytoskeleton research. In this volume we have emphasized the techniques commonly used in modern cytoskeletal research. Clearly, the researcher in this field needs to be conversant with a variety of approaches, though not necessarily expert in all of them. Recognizing the varied interests and capabilities of different laboratories, we have tried to provide some variation in the levels of sophistication in the descriptions of more complex technologies. In this manner we hope that the volume will be of value to both the uninitiated and the expert in the various aspects of cytoskeletal research. Finally, a multitude of systems are being exploited in cytoskeletal research. Since describing even a fraction of these would be an encyclopedic undertaking, we have tended to emphasize general approaches except where specific examples illuminate these general approaches.

Miami   K. L. C.
May 1991   C. A. C. C.

Contents

3. Actin filament assembly and organization in *vitro* 47

John A. Cooper

4. Mapping structural and functional domains in actin-binding proteins 73

Paul Matsudaira

Contents

5. The generation and isolation of mutant actins

R. Kimberley Cook and Peter A. Rubenstein

9. Methods for studying the cytoskeleton in yeast   197

*F. Solomon, L. Connell, D. Kirkpatrick, V. Praitis,
and B. Weinstein*

Contributors

CORALIE A. CAROTHERS CARRAWAY
Department of Biochemistry and Molecular Biology, University of Miami School of Medicine, Miami, FL 33101, USA.

R. KIMBERLEY COOK
Department of Biochemistry, University of Iowa College of Medicine, Iowa City, IA 52242, USA.

JOHN A. COOPER
Department of Cell Biology and Physiology, Washington University School of Medicine, St. Louis, MO 63110, USA.

L. CONNELL
Department of Biology, MIT Building E–17, Cambridge, MA 02139, USA.

ALBERTO DOMINGO
Molecular, Cellular, and Developmental Biology, University of Colorado, Boulder, CO 80309–0347, USA.

ROBERT M. EVANS
Department of Pathology, University of Colorado Health Sciences Center, Denver, CO 80262–0216, USA.

JOHN H. HARTWIG
Experimental Medicine, Brigham and Women's Hospital, Harvard Medical School, Longwood Medical Research Center, 221 Longwood Avenue, Boston, MA 02115, USA.

D. KIRKPATRICK
Department of Biology, MIT Building E–17, Cambridge, MA 02139, USA.

MICHAEL W. KLYMKOSWKY
Molecular, Cellular, and Developmental Biology, University of Colorado, Boulder, CO 80309–0347, USA.

PAUL MATSUDAIRA
Whitehead Institute for Biomedical Research and Department of Biology, MIT, Nine Cambridge Center, Cambridge, MA 02142, USA.

V. PRAITIS
Department of Biology, MIT Building E–17, Cambridge, MA 02139, USA.

Contributors

PETER A. RUBENSTEIN
Department of Biochemistry, University of Iowa College of Medicine, Iowa City, IA 52242, USA.

ALFONSO J. SARRIA
Department of Pathology, University of Colorado Health Sciences Center, Denver, CO 80262–0216, USA.

ROGER D. SLOBODA
Department of Biological Sciences, Dartmouth College, Hanover, NH 03755, USA and The Marine Biological Laboratory, Woods Hole, MA 02543, USA.

F. SOLOMON
Department of Biology, MIT Building E–17, Cambridge, MA 02139, USA.

YU-LI WANG
Cell Biology Group, Worcester Foundation for Experimental Biology, 222 Maple Avenue, Shrewsbury, MA 01545, USA.

ROBLEY C. WILLIAMS, JR
Department of Molecular Biology, Vanderbilt University, Nashville, TN 37235, USA.

B. WEINSTEIN
Department of Biology, MIT Building E–17, Cambridge, MA 02139, USA.

Abbreviations

AMP-PNP	adenosine $5'(\beta, \gamma\text{-imino})$triphosphate
AP	acrosomal process
AP	alkaline phosphatase
BCA	bicinchoninic acid
BCIP	disodium p-nitrophenyl phosphate
BNPS-skatole	2-(2'-nitrophenylsulphenyl)-3-methyl-3-bromoindolenine
BSA	bovine serum albumin
CAPS	3-(cyclohexylamino)-1-propanesulphonic acid
CHEF	contour-clamped homogeneous electric field
Con A	concanavalin A
CSAT	cell-surface antigen, one of the integrins
DABCO	1,4-diazabicyclo[2.2.2]octane
DAPI	4',6'-diamidino-2-phenylindole dihydrochloride
DC	direct current
DDE	demembranation-dynein extraction
DEAE	diethylaminoethyl
DCD	diethylene glycol distearate
DIC	differential interference contrast
DME	Dulbecco's modified Eagle's medium
DMF	dimethylformamide
DMSO	dimethylsulphoxide
DTE	dithioerythreitol
DTT	dithiothreitol
EGTA	ethylene glycol-bis(β-aminoethyl ether)N,N,N',N'-tetra-acetic acid
ELISA	enzyme-linked immunosorbent assay
FACS	fluorescence-activated cell sorter
FPLC	fast protein liquid chromatrography
FPR	fluorescence photobleaching recovery
FRAP	fluorescence recovery after photobleaching
5-FOA	5-fluoroorotic acid
GFAP	glial fibrillar acidic protein
GSD	glycerol-SDS-dithiothreitol
GTP	guanosine-5'-triphosphate
Hepes	N-(2-hydroxyethyl)piperazine-N'-(2-ethanesulphonic acid)
HRP	horse radish peroxidase
ICCD	intensified charge-coupled-device
IF	intermediate filament
IFsp	intermediate filament subunit protein
IgG	immunoglobulin G

ISIT	intensified silicon-intensified-target
KLH	keyhole limpet haemocyanin
MAP	microtubule-associated protein
MBS	maleimidobenzoyl-N-hydroxysuccinimide ester
MCP	macrophage capping protein
MES	2-[N-morpholino]ethane sulphonic acid
MMTV	mouse mammary tumour virus
NA	numerical aperture
NBD	7-chloro-4-nitro-benzooxadiazole
NBT	nitro blue tetrazolium
NF	neurofilament
nm	namometre
NP-40	Nonidet P-40
NTCB	2-nitro-5-thiocyanobenzoic acid
OMDR	optical memory disk recorder
PAGE	polyacrylamide gel electrophoresis
PBS	phosphate-buffered saline
PCIA	phenol:chloroform:isoamyl alcohol, 25:24:1
PCR	polymerase chain reaction
PEG	polyethylene glycol
PHEM	Pipes–Hepes–EGTA–$MgCl_2$
Pipes	piperazine-N,N'-bis(2-ethanesulphonic acid)
PMEG	Pipes–Mg–EGTA–GTP buffer
PMSF	phenylmethyl sulphonyl fluoride
PVDF	polyvinylidene fluoride
RF	replicative form
RSV	Rous sarcoma virus
SCE	sorbitol–sodium citrate–EDTA
SDS	sodium dodecyl sulphate
SKEM	sucrose–KCl–EGTA–Mg–imidazole buffer
SMCC	succimidyl 4-(n-maleimidomethyl)cyclohexane-1-carboxylate)
ssDNA	single-stranded DNA
TE	Tris–EDTA buffer
TFA	trifluoroacetic acid
TIU	trypsin inhibitor units
TPCK	tosylphenylchloromethyl ketone
WGA	wheat germ agglutinin
YPD	yeast extract–peptone–dextrose

Fluorescence microscopic analysis of cytoskeletal organization and dynamics

YU-LI WANG

1. Introduction

Fluorescence microscopy represents a simple, efficient approach for examining the distribution of specific molecules in cells. Such spatial information is particularly important for the study of the cytoskeleton, since the molecular components involved often undergo dramatic reorganizations in response to internal and external stimuli. During the past two decades fluorescence microscopy has served as a bridge between biochemical characterization and cellular functions, and has provided important clues to the mechanisms of various motile activities and the functional roles of many cytoskeletal proteins.

Most applications of fluorescence microscopy to date involve fixing cells and staining with either specific antibodies (immunofluorescence) or specific fluorescent ligands such as phalloidin. The application of multiple fluorophores has allowed up to five separate components to be mapped and correlated within a given cell. The main limitation of this approach is that all observations are made with fixed cells, thus it can be difficult to reconstruct dynamic events based on still images of different cells. The second approach, fluorescent analogue cytochemistry, is based on the observation of functional, fluorescently-labelled proteins microinjected into living cells. It allows not only the direct observation of specific components during dynamic processes, but also the analysis of protein interactions and molecular mobilities.

Laboratory procedures for both fluorescent staining and fluorescent analogue cytochemistry have been discussed in numerous articles (1–7). The purpose of the present article is to describe our current procedures for the fluorescence microscopy of cultured animal cells, and to provide additional details that are not already available.

2. Fluorescent staining of fixed, permeabilized cells

Fluorescent staining, including troubleshooting and potential artefacts, has been discussed in the article by K. Wang *et al.* (1). Thus, this section describes only briefly our current methods for indirect immunofluorescence and for staining with fluorescent phalloidin.

The fixation procedure should be chosen on the basis of the specific purpose of the experiment. For example, the glutaraldehyde–Triton protocol developed by Small (1, 8) results in a superb preservation of lamellipodia. However, glutaraldehyde induces fluorogenic side reactions; the resulting background fluorescence can be reduced but not totally removed by treatment with sodium borohydride. The high degree of cross-linking by glutaraldehyde may also destroy some antigenic sites for immunostaining. Fixation–extraction with -20 °C methanol containing 5–50 mM EGTA is also commonly used for microtubule staining (2). However, this method does not preserve lamellipodia and also interferes with the staining of actin filaments by fluorescent phalloidin. On the other hand, formaldehyde usually offers satisfactory results for microfilament structures but is known to disrupt microtubules (2).

The fixation method that we typically use for studying the microfilament system is described in *Protocol 1*. A variation of the method is to include 0.1–0.5 per cent Triton-X 100 in the fixation solution, and delete the acetone extraction step. The cover slip is rinsed three times with PBS after fixation for 5–10 min. This method results in a faster fixation due to the permeabilization by Triton, and a better preservation of the lamellipodia. However, the use of Triton also results in an extensive extraction of soluble contents, including microinjected soluble molecules, from the cell. Cold acetone on the other hand removes primarily lipids while precipitating and retaining most proteins. Other permeabilizing agents, such as saponin, may also be used.

Protocol 1. Fixation and extraction of cells

Materials

- Phosphate buffered saline (PBS): 137 mM NaCl, 2.7 mM KCl, 8 mM Na_2HPO_4, 1.5 mM KH_2PO_4.

- Formaldehyde, 4 per cent, in the PHEM buffer (9): 60 mM Pipes, 25 mM Hepes, 10 mM EGTA, 3 mM $MgCl_2$ at pH 6.1. The quality of formaldehyde is critical. We use the 16 per cent 'Electron miscroscope' grade stock solution from Electron Microscopy Sciences, supplied as aliquots in ampoules. Consistent results can also be obtained with paraformaldehyde, prepared as described by Wang *et al.* (1).

2

• Acetone (analytical grade), chilled to −20 °C: Acetone can be chilled either in an explosion-proof freezer, or in a dry ice–acetone bath and then allowed to sit at room temperature for about 5 min. Precise temperature control of acetone is not necessary but a temperature below −20 °C should be avoided as it leads to formation of ice on the surface of the cover slip and resulting distortion of cell morphology.

Method

1. Grow culture cells on substrates that are free of autofluorescence (most plastic substrates are not suitable), usually glass cover slips.
2. Rinse the culture 2–3 times with warm (37 °C) PBS to remove serum proteins and other components in the medium that may interfere with fixation.
3. Immediately drop the cover slip into warm, freshly prepared formaldehyde solution. Fix for 10 min.
4. Rinse the cover slip twice with PBS at room temp.
5. Dip the cover slip into −20 °C acetone for 5 min.
6. Allow the cover slip to air dry at room temp.

A typical method for staining cells with fluorescent phalloidin or antibodies is described in *Protocol 2*. The most important consideration for antibody staining is the quality of the antibodies and the proper dilution of the antibody solution. The primary antibody should be tested carefully by Western blot analysis of cell lysates. Reliable fluorescent secondary antibodies can usually be obtained from commercial sources (e.g. Sigma Chemical, Organon Teknika-Cappel, or TAGO Inc.), and fluorescent conjugation in the lab is generally unnecessary. The dilution factor should be determined empirically by staining multiple samples with different concentrations. The lowest concentration that gives detectable signals should be used.

Protocol 2. Staining cells with fluorescent antibodies or phalloidin

Materials

• PBS
• PBS with 1 per cent bovine serum albumin, filtered through filter paper. Sodium azide (0.02 per cent) may be included to inhibit microbial growth.
• Fluorescent phalloidin, 50–100 nM in PBS: Fluorescent phalloidin is supplied by Molecular Probes as a stock solution in methanol, and is usually used after drying in a test-tube and redissolving in PBS with thorough vortexing. The stock solution can also be directly diluted into

Protocol 2. *Continued*

PBS; however, it is important to make sure that the final concentration of methanol is below 5 per cent.

• Primary and secondary antibodies, diluted in PBS–BSA and clarified in a table-top centrifuge for 10–20 min.

Methods

1. Incubate the cover slip from *Protocol 1* with primary antibodies. The required volume of antibody solution can be reduced by cutting a piece of parafilm approximately the area to be stained, and placing a drop (50–100 μl) of the antibody solution on the parafilm. The cover slip is then inverted on to the solution, which spreads uniformly over the surface. It is important to keep the cover slip from drying out after the application of the primary antibody.

2. Place the cover slip face down in a moist chamber (e.g. a closed Petri dish containing a piece of wet paper tissue) and incubate for 45–120 min at 37 °C. Alternatively, the staining can be allowed to proceed overnight at 4 °C.

3. Invert the cover slip (face up) and place in a dish containing PBS/BSA at room temp. The parafilm usually floats off automatically.

4. Agitate the container gently on a shaker table for 5 min.

5. Repeat the washing procedure twice more.

6. Apply the fluorescent secondary antibody as for the primary antibody.

7. Staining with fluorescent phalloidin can be performed, after antibody staining, by incubating the cover slip with 50–100 nM fluorescent phalloidin in PBS for 20 min at room temp., followed by 2–3 rinses with PBS before observation.

The cover slips can be observed after mounting in PBS. However, special 'anti-fading' mounting solutions are commonly used to reduce the photobleaching during observation (2, 10). One solution suitable for long-term preservation of the sample is prepared by dissolving 100 mg/ml 1,4-diazabicyclo [2,2,2] octane in Mowiol mounting medium, as described by Osborn and Weber (2). A solution without Mowiol works in the short term, and is much easier to prepare. In this case the cover slip can be sealed with nail polish around the edge and stored at 4 °C for several days.

Detailed discussions of the potential problems of immunofluorescence and important controls for staining can be found in Wang *et al.* (1). A problem that can particularly affect staining with monoclonal antibodies is the availability of antigenic sites, which can be affected by fixation, extraction, protein conformation, or protein–protein interactions. Thus, either the entire cell or a part of it may not show staining even though the antibody is functionally active. A survey of different fixation and extraction methods as well as different antibody preparations should be performed if this is in question. In addition, treatment of fixed cells with trypsin or SDS may aid in the exposure of shielded antigenic sites. Although the differential reactivities of monoclonal antibodies generally represent a problem in immuno-fluorescence, such properties may sometimes be used as a powerful tool for probing the conformational states of molecules in the cell (11).

Another potential problem with fluorescent staining is the redistribution of proteins during fixation. One scenario involves proteins that interact weakly with cytoskeletal structures. Fixation may cause such interactions to become irreversible and result in a significant increase in the association with cytoskeleton (12). Although different fixation methods may be tested, it can be difficult to decide which one to trust if different results are obtained. Thus while immunofluorescence is useful for determining the presence of specific proteins on cytoskeletal structures, the degree of association should be interpreted with caution. The use of fluorescent analogue cytochemistry, as discussed below, represents an independent approach for the localization of specific proteins.

3. Fluorescent analogue cytochemistry

Fluorescent analogue cytochemistry involves the purification, fluorescent labelling, and microinjection of specific proteins into living cells. The approach is based on the assumption that microinjected analogues behave similarly to the endogenous counterpart, and can thus provide information about the physiological behaviour of cellular components based on the fluorescence signal. The major advantage of fluorescent analogue cyto-chemistry over immunofluorescence is its ability to yield direct information on dynamic changes and even molecular interactions within living cells. The reader is referred to a recent review article for a detailed discussion of previous applications and future potentials of fluorescent analogue cyto-chemistry (13).

3.1 Preparation of fluorescent analogues

Several detailed protocols for the fluorescent labelling of cytoskeletal components have been published previously (*Table 1*), so the discussion here emphasizes general principles. Proteins to be conjugated should be prepared in at least milligram quantities. Although only a small amount of protein (< 50 µg) is required for microinjection, a minimum of about 0.5 mg is usually necessary for fluorescent labelling. Other important considerations for the choice of proteins have been discussed in Wang (7). For most morphological studies, members of the rhodamine family, such as tetra-methylrhodamine, provide relatively stable and strong signals with minimal interference from autofluorescence. The choice of fluorophores is more critical for fluorescent analogue cytochemistry than for immunofluorescence (4), particularly since anti-bleaching agents cannot be applied to living cells, and strong excitation light can easily cause adverse cellular reactions. Many fluorophores are now available in forms reactive toward either amino or sulphydryl groups. The reagent should be reactive under mild buffer conditions, and form stable, covalent bonds with minimal effects on the function of the protein. Although it is difficult to predict an optimal reactive group, often a number of fluorescent dyes can be used with equally satisfactory results. It is strongly recommended that a 'Handbook' be obtained from Molecular Probes, since it contains valuable information about the physical and chemical properties of a large number of fluorescent reagents. In addition, the book by Means and Feeney on *Chemical modifications of proteins* (14) should be consulted for the characteristics of many common labelling reactions.

Proteins are dialysed into an appropriate reaction buffer at a concentration of 1.0–10.0 mg/ml. Components that may interfere with the labelling reaction, including Tris, dithiothreitol (DTT), and ATP, should be removed or reduced in concentration during dialysis. If the cysteine thiol groups are to be labelled, it is recommended that 5–10 mM DTT be added before dialysis to recover any oxidized thiol groups (unless the protein contains structural disulphide bridges). The protein is then dialysed overnight to remove the added DTT.

Fluorescent reagents should be prepared fresh and are usually dissolved in an aqueous buffer before adding to the protein solution. A molar excess of 10 to 20 over the protein represents a reasonable starting point, and can be adjusted up or down based on the resulting stoichiometry of labelling. Since many reactions involve the release or incorporation of protons, it is important

Table 1. Some published protocols for the preparation of fluorescent analogues of cytoskeletal proteins

Protein	Fluorophore	Reference
Actin	Fluorescein	15
Actin	Lissamine rhodamine B	5
Myosin	Tetramethylrhodamine	35
Alpha-actinin	Tetramethylrhodamine	1
Alpha-actinin	Fluorescein	4
Alpha-actinin	Fluorescein and lissamine rhodamine B	5
Vinculin	Fluorescein and lissamine rhodamine B	5
Tubulin	Fluorescein	6

to use a buffer of sufficient strength for maintaining pH. The buffering capacity can be tested by adding the estimated amount of acid or base and measuring the change in pH. We commonly use 50–100 mM borate or Hepes for reactions that require a slightly alkaline pH (8.0–9.0), and phosphate or Pipes for pH values near neutrality.

Of prime importance is the technique for solubilizing the reagent, since variable degrees of solubilization can lead to highly inconsistent labelling. Many reagents are poorly soluble in water and may require hours to dissolve if added directly to an aqueous buffer. The solubilization process is often facilitated by the use of a small amount of organic solvent such as acetone, dimethylsulphoxide (DMSO), dimethylformamide, or ethanol to form a stock solution or fine slurry, which is then added very slowly into the aqueous labelling buffer. Ths solubility of some compounds, such as fluorescein-containing reagents, is also significantly affected by the pH of the buffer.

Incomplete solubilization can be difficult to detect. When in doubt the solution should be clarified in an ultracentrifuge before adding to the protein. The consistency of reaction with such partially soluble reagents can be improved by measuring the absorbance of the supernatant after centrifugation and adjusting the volume used for the reaction. Alternatively, some poorly soluble reagents such as lissamine rhodamine sulphonyl chloride are available pre-absorbed on celite, and can be gradually released after adding directly to the protein solution. Celite particles are then removed by centrifugation at the end of reaction. A simple variation of this method is to dissolve the reagent in a volatile solvent and dry on to the side and bottom of reaction vessels; protein solutions are then added to initiate the reaction.

The optimal reaction time and temperature vary with proteins and reagents and should be determined empirically. A period of 2 h at 4 °C can be used as a starting point. Unless celite particles are used, constant stirring is not recommended after the initial mixing. The reaction should be terminated with quenchers such as lysine, cysteine, or DTT, at molar ratios of 5–10 times over the reagent. The quenching not only makes the reaction more reproducible, but can also facilitate the removal of unreacted non-polar reagents.

Various methods have been used for the removal of unreacted reagents, including dialysis, gel filtration chromatography, adsorption chromatography, and ion exchange chromatography (7). In addition, cycles of precipitation and solubilization can effectively remove a large fraction of unreacted dyes. For proteins with the ability to self-assemble, cycles of polymerization and depolymerization serve both to remove unreacted dyes and to select for assembly-competent molecules.

Purified fluorescent proteins are concentrated to 1–5 mg/ml, usually with an Amicon Centricon concentrator or by vacuum dialysis in Collodion bags (Schleicher and Schuell). The Centricon concentrator is faster, but may become clogged with some proteins. Before microinjection, labelled proteins should be examined by SDS-PAGE to make sure that there is no detectable

7

residual free dye, which may still be present even after an apparently effective gel filtration step. The fluorophore:protein ratio is determined based on the absorbance of the fluorophore, the estimated extinction coefficient of the fluorophore, and a standard protein assay such as the Lowry or Bradford assay. We avoid the use of absorbance at 280 or 290 nm for the determination of protein concentration, since most fluorophores absorb sufficient UV light to cause considerable errors. The major source of error in measuring the labelling stoichiometry usually comes from the extinction coefficient of the fluorophore, which can vary significantly following the conjugation to proteins. A simple method for the estimation of the extinction coefficient is described in Wang and Taylor (15). An apparent labelling ratio of 1–2 fluorophores per polypeptide is adequate for most experiments. In addition to labelling stoichiometry, possible effects of fluorescent labelling on the function of the protein should be determined based on the known biochemical properties. For experiments involving quantitative fluorimetry, it is also necessary to characterize the spectroscopic properties of the conjugate, such as possible changes in the emission spectrum, as a function of conformation.

The fluorescent analogue is then dialysed into a microinjection buffer. We have successfully used as a carrier solution 5–10 mM Tris–acetate at pH 6.95– 7.05, although others have used a variety of buffers for microinjections (e.g. 16). The protein solution should be clarified in an ultracentrifuge (7).

3.2 Cell culture and microinjection

The main concern of cell culture is to maintain the viability of the cell during microinjection and observation, and to provide an optimal optical quality for fluorescence microscopy. We choose to use a stage incubator on an inverted microscope and a relatively simple cover slip chamber/dish (17). Alternatively, some investigators culture cells on cover slips placed inside 60 mm cell culture dishes, and subsequently remove the cover slips and mount in observation chambers. Various approaches have been discussed in McKenna and Wang (17). Depending on the cell type, the cover slip can be used without coating or after coating with polylysine or extracellular matrix proteins such as the ECL matrix (Upstate Biotechnology).

It is critical to maintain the cells under optimal conditions. Inadequately maintained cells may appear normal in phase optics, but recover poorly after microinjection. Different media vary considerably in their level of fluorescence, which can be a problem when signals from microinjected cells are extremely low. The fluorescence of the medium comes mainly from NADH, riboflavins, and phenol red (18) which are present at a relatively high level in Dulbecco's modified Eagle's medium (DME). A method used in this laboratory is to maintain stocks of cells in DME but plate cells in F12K medium for microinjection. For some experiments it may be advantageous to plate cells in Leibovitz L-15 medium, which uses phosphate buffer and does not require carbon dioxide for the maintenance of pH. However, the medium for maintaining stocks of cell lines should not be changed arbitrarily from that originally used for raising the parent cell line.

Table 2 lists special equipment and supplies used for microinjection. Microneedles are drawn from glass capillaries, using pipette pullers that are available from manufacturers such as David Kopf. Elaborate pulling-control mechanisms are unnecessary for preparing needles for microinjection; however, it is critical to be able to adjust the shape of the needle by changing the current of the heating element and the pulling force. Needles with long tapers tend to trap bubbles, and those with short tapers are usually too fragile at their tips.

Table 2. Special items required for microinjection

Ultracentrifuge and rotors for small volumes
Protein concentrator
Microdialysis system
Vibration-free table
Inverted microscope with phase and epifluorescence optics
Diamond scribe
Microscope stage heater
Microscope stage carbon dioxide outlet
Microcapillaries
Drawn-out Pasteur pipettes
Microneedle puller
Microneedle holder
Air pressure regulator (e.g. syringe) and tubing connections
Micromanipulator

We use capillaries containing inner threads (Frederick Haer) for the preparation of needles. The inner thread allows the needle to be loaded from the back end with a drawn-out pipette. The solution automatically flows into the tip within a few seconds due to capillary action. The glass capillaries from Frederick Haer are used without washing on treatment, except in special cases where the external surface of the capillary is siliconized before pulling to minimize the build-up of materials at the tip of the needle.

Microinjection can be performed conveniently for most cultured animal cells with 25–40× phase objectives and 10× eyepieces. High-quality phase optics are required for detailed observations during microinjections. The microscope and the micromanipulator should be isolated from vibration (7), so that no visible movement of the needle is detectable under the microscope. The micromanipulator should be equipped with coarse controls for tilting and for movement in the $x-y$ direction, and fine controls for movement along the z axis. Other features such as joysticks are useful but not critical. A simple, functional micromanipulator can be assembled with optical positioning devices for under US$1500 (e.g. supplied by the Newport Corp.).

9

Pressure for microinjection can be simply controlled with an empty 10–50 ml syringe (1, 19). However, some experience is required to maintain the pressure in an optimal range, typically between 0.25 and 0.75 p.s.i. Alternatively, simple electronic or mechanical devices can be designed to allow more precise and reproducible controls of the pressure. The rate of flow during microinjection is affected by the applied pressure, the size of the tip, the viscosity of the material, the amount of materials accumulating at the tip, the presence of any air bubbles in the needle, and the resistance of the cell. Thus the pressure must be adjusted according to the actual response of the cell.

There are also commercial devices for delivering pulses of solution into cells. Although this method reduces to some extent the variability in the volume of microinjection, in our hands short, strong pulses tend to cause more severe shocks to the cell than do gentle, sustained flows. In addition, because of the sensitivity of the rate of injection to the multiple factors mentioned above, it is unrealistic to assume that identical volumes can be microinjected into each cell. In practice, the precision of microinjection experiments is also limited by the variability of cell volume with most cell lines.

Our method of microinjection, described in detail in *Protocol 3*, is derived from that of Graessman *et al.* (19). In this approach, microinjection is achieved by positioning cells under the needle and dipping the needle into the cell, while maintaining the needle at a relatively constant position in the X–Y plane. The flow is maintained constantly at a relatively low level and the amount of microinjection is controlled by the duration that the needle stays inside the cell. Gentle microinjections induce a transient decrease in the phase density of the injected region, and an increase in the contrast of the nucleus against the cytoplasm. In addition, cells may show transient responses to the microinjection, notably the retraction of lamellipodia, but should recover completely within 30 min. Unless the experiment requires an immediate observation, it is best to incubate cells for at least 1 h in a regular incubator following the microinjection to allow complete recovery and incorporation of molecules. For prolonged observations, it is critical to keep the cells in a microscope incubator or perfusion chamber, in order to maintain a proper temperature, osmolarity, and pH (17).

Protocol 3. Microinjection

1. Plate the cells (17) and allow them to recover for 12–48 h.

2. Scan the dish at a low magnification to choose a good area for microinjection. The cells should be well attached at this time, otherwise they may dissociate from the dish during microinjection. For most experiments, the cell density should be somewhat below confluency to

10

provide an adequate space for cells to spread and for manoeuvring the microinjection needles.

3. Gently mark areas to be microinjected with a diamond-tipped scribe on the exterior surface of the cover slip.

4. Locate regions marked for microinjection and bring cells into focus. Place a nearby cell-free area in the centre of the field; use 250–400 ×.

5. Draw microneedles with a pipette puller. Needles may be prepared beforehand and stored in a dust-free container. However, prolonged storage may affect sterility of the needle and is not recommended.

6. Load a microneedle. When capillaries containing inner threads are used, the solution is loaded from the back end of the needle near the tapered region using a drawn-out Pasteur pipette.

7. Mount the microneedle on the needle holder and the micromanipulator.

8. Carefully introduce the needle into the medium, but keep the needle at a level well above cells. Apply pressure to prevent backflow of the medium into the needle. Bring the needle near the centre of the field. The needle should be out of focus in the microscope, appearing only as a shadow which can be discerned most easily by moving the needle back and forth with the coarse X–Y control of the micromanipulator.

9. Bring the needle close to focus but still above the level of the cells. Position a cell to be microinjected directly under the needle. For well spread cells, microinjections have to be performed in the thicker, perinuclear region.

10. Carefully bring the needle to focus, i.e., lower the needle down into the cell. Microinjection sometimes starts as soon as the needle is brought into focus. Otherwise a gentle tap on the micromanipulator is used to puncture the cell membrane.

11. Terminate the microinjection by raising the needle. Adjust the injection pressure if necessary. Move the next cell under the needle, and repeat the microinjection.

12. Replace the medium after microinjection, to correct for any evaporation that may have occurred during microinjection and to remove materials expelled from the needle into the medium.

13. Incubate cells for at least 1 h in a regular incubator.

The volume of microinjection can be calculated from the integrated intensity of fluorescent molecules injected into the cell. A method derived from Lee (20) is described in detail in *Protocol 4*. Typically the pressure microinjection technique allows the volume of delivery to vary between 0.5 and 10 per cent of the cell volume. However, in a particular experiment, the volume can be maintained within a factor of 2–3 by consistent performance of the injection technique.

Protocol 4. Measurement of microinjection volume

Materials

- Lucifer Yellow–dextran (10 kDa, Molecular Probes); 2 mg/ml and 10 mg/ml in 5 mM Tris–acetate, pH 7.0.

Methods

1. Microinject cells with 10 mg/ml Lucifer Yellow–dextran.
2. Measure the total fluorescence intensity from each injected cell.
3. Measure the background intensity and subtract from the signal obtained in the previous step. Background image can be obtained after defocusing the microscope. Alternatively, background intensity can be determined by measurement at several spots in the vicinity of the injected cell and multiplying the average by the number of pixels covered by the cell.
4. Perform shading correction if high accuracy is required.
5. Prepare an observation vessel containing type FF immersion oil (Cargille Laboratories), 2–4 mm in depth.
6. Load 10–20 µl of 2 mg/ml Lucifer Yellow–dextran into an atomizer, which can be constructed by attaching tubing to a dusting spray can.
7. Spray the solution of Lucifer Yellow–dextran into the immersion oil.
8. Observe the droplets of dextran solution under the microscope. Obtain the integrated fluorescence intensity from droplets of different sizes. Also measure the diameter of each droplet.
9. Plot the droplet fluorescence intensity against volume.
10. Find the droplet volume that yields the same fluorescence intensity as that measured from the cell. Divide this volume by 5 to obtain the volume of microinjection.
11. Measure the average cell volume by inducing cell rounding (e.g. by gently trypsinizing the culture) and measuring the average diameter. This is used for the calculation of injection volume relative to the cell volume.

3.3 Limitations and potential problems

As with other methods for living cells, microinjections of fluorescent analogues may be complicated by the unpredictable nature of living cells and perturbations caused by the physical manipulation. In addition, the success of the approach may be affected by the possible modification of the injected analogues in the cell and potential problems of incorporation into physiological structures. The problem of cellular disruption is usually analysed by examining cellular activities such as locomotion and division, and cellular structures based on immunofluorescence staining or ultrastructural analyses. The issue of incorporation is usually addressed by performing immunofluorescence microscopy to compare the distribution of the injected analogue with that of the endogenous counterpart. The two should appear similar if not identical to each other throughout the course of any process under study. However, as immunofluorescence may suffer from its own artefacts, a discrepency between the two does not necessarily indicate a problem with the incorporation of analogues.

4. Image detection and processing

The use of low-light-level detectors and image-processing systems has revolutionized the analysis of fluorescence signals from single cells. These powerful devices not only play a critical role in the detection of extremely weak signals of microinjected fluorescent analogues, but also greatly facilitate the analysis of fluorescence staining of fixed cells (*Table 3* lists equipment required). For a comprehensive discussion of the technical aspects of video microscopy, the book by Inoue (21) is an excellent reference and contains many important details that are not covered here. The following discussion will emphasize the detection of signals in fluorescent analogue cytochemistry.

As discussed previously (21, 22), fluorescence should be observed with objective lenses of high numerical apertures. The lenses most commonly used in this lab are 25× Plan-Neofluar/NA 0.8, 63× Neofluar/NA 1.25, and 100× Neofluar/NA 1.30 objectives. A great degree of flexibility in magnification is gained by using a coupling system between the microscope and the detector that allows changes of projection lenses (21). The level of excitation light for fluorescence observation should be maintained as low as possible, to avoid both photobleaching and radiation damage to living cells. For the detection of fluorescent analogues, it is also important to minimize the number of images recorded by adjusting the frequency of image recording according to the rate of the event under study. Proper heat filtration in the excitation light path is crucial for minimizing the heating of cells and removing infrared light that can be picked up by many video detectors.

The detection of images represents a special challenge in fluorescent analogue cytochemistry, since the level of signals is limited both by the number of molecules that can be incorporated and by the low excitation light that can be used with live cells. As a result, the required sensitivity of detection is generally higher than that for immunofluroescence or ion imaging. A simple guideline is that the detector should be sensitive enough to pick up images

Table 3. Equipment for low-light-level image acquisition and processing

Items	Considerations
Optical coupling between microscope and detectors	Ability to change magnification
	Light throughput
Detector	Sensitivity
	Resolution
	Response time
	Geometric distortion and linearity
	Requirement of image-processing systems
	Communication with computer
Image-processing hardware	Real-time processing of video frames
	Number of frame buffers
	Resolution of frame buffers
	Bit-width of frame buffers
	Look-up tables
	Speed of calculation and input/output
	Speed of communication with the computer system
Computer system	Speed of calculation and input/output
	Number of communication ports
	Backup systems (e.g. streaming tape)
	Maintenance and supervision required
Computer software	Availability of key functions
	Ease of use
	Efficiency of programs
Hard-disk drives	Capacity
	Speed
Pointing devices	Resolution
	Response time
	Ability to draw precisely
	Ease of pointing/defining specific points
Monitors	Resolution
	Range of contrast and brightness
	DC restoration circuit

Optional equipment
Image printers
Video recorders

barely visible to the dark-adapted naked eye. This can be achieved with intensified SIT (ISIT) or intensified CCD (ICCD) video cameras, or with a cooled, slow-scan CCD detector commonly used in astronomy. Detailed discussions of the characteristics of various detectors can be found in several articles (23–26).

As the signal level decreases, more experience is needed to bring raw images into focus with ISIT or ICCD cameras. In some cases it is helpful to perform a 'rolling average' with an image processor to improve the image quality during focusing. However, this also causes a significant delay in response time and makes it necessary to focus very slowly. The problem is much more serious with slow-scan CCD detectors, which may need an accumulation time of 0.5–2 s to yield even low-resolution images for focusing. The only satisfactory solution so far is to place the CCD detector parfocal with an ISIT/ICCD camera at two separate ports, and to use the video camera for focusing.

In order to obtain images of high resolution, it is critical to perform signal integration to improve the signal/noise ratio of low-light-level images. This is typically achieved with an image processor for video cameras and a timing circuit for slow-scan CCD detectors. Since the signal/noise ratio is proportional to the square root of the integration time, the improvement is most noticeable during the first few seconds and levels off subsequently. With ISIT video cameras, an integration of 2–10 s represents a compromise between signal/noise ratio and integration time. An integration time of 5–20 s is typically used with a cooled slow-scan CCD detector (Star I; Photometrics). In our experience little can be gained with a longer integration if the signal is too low to yield a usable image with these integration times. At such low levels of signal, stray light and the fluorescence of the medium often become the limiting factors, raw images may be almost impossible to focus, and focusing may also show a noticeable drift during prolonged integrations.

Background images, which are obtained by switching off the excitation light and integrating for the same period of time, are usually subtracted from integrated fluorescence images. The subtraction can be done either before or after appropriate division (usually by the number of integrated frames from a video camera) before viewing in a frame buffer. When the signal level is very low, it is better to subtract the integrated background first. The results are then divided in a stepwise fashion until no overflowing pixels are observed in the frame buffer. This method minimizes the loss of information associated with division and rounding, and has been used successfully to image structures that are barely above the noise level (27).

The images are best stored as digital files, both for preservation of information and for ease of analysis, as discussed later. Simple programs that automatically generate a series of file names, and imprint the time and date on the image or image files, can greatly improve productivity. The only drawback of digital files is that a lot of computer disk space is required— about 0.25 Mb for a $512 \times 480 \times 8$ bit image. We have alleviated this problem by compressing the image files. Briefly, since most fluorescence images are characterized by a large number of pixels of similar intensities and gradual transitions in intensities, a program can be used to convert the pixel intensities into differences between neighbouring pixels. The differences are then coded into variable bits, such that the most frequent difference (i.e. 0) is stored only as a single bit and infrequent differences stored with up to 12 bits (28). A 2–3 fold compression can be obtained using this approach, typically

15

within 5 s on a graphics workstation. However, the ideal solution to the problem of image storage is the use of removable, optical read–write disks of gigabyte capacities, which are now commercially available.

5. Analysis of fluorescence images

Images of fluorescence microscopy are usually analysed in terms of the distribution of fluorescence intensities and/or changes in fluorescence as a function of time. Both qualitative and quantitative analyses can be greatly facilitated by the use of computers. Discussed below are several useful techniques that can be readily implemented into most image-processing systems.

Table 4. Useful computer programs for image processing

Programs	Considerations
1. Basic functions for image acquisition and processing	
Average frames	Handling of overflow
Accumulate frames on frame buffers	Bit-width for the accumulated frame
Take rolling average	
Subtract two frames	Correction of negative values
	Bit-width
Divide two frames	Speed
	Handling of scaling factors
Divide images by a constant	Speed when dividing by power of 2
	Ability to divide by a decimal number
Store images on files	Speed
View image files	Ability to compress/decompress during storage/retrieval
View images in composite form	Ability to load/store images in specific frame buffers
	Ability to load a series of images automatically
Stretch contrast	Ease to specify and change the extent of stretching
	Ease to remove the stretching
Show motion picture	Ability to change speed
Alternate frame buffers for display	Ability to stop, step, and reverse direction
	Ability to roam images for correcting misregistration
2. Basic functions for image analysis	
Define areas of interest	Ability to define lines, rectangular regions, and regions of irregular shapes
	Ease of moving or editing defined areas

16

Measure distances	Ability to convert into actual distances according to the magnification
	Ease of performing statistics
Measure point intensities	Ease of specifying random points
	Ability to perform averaging over local areas
Integrate intensities over an area	Ability to integrate over irregular areas
	Display of number of pixels within the defined area
Display intensity profile	Ability to perform averaging over local areas
3. Miscellaneous functions	
Remove noise	Minimal effect on resolution
Convolute	Limitation on the size of the matrix
	Ability to normalize results when the sum of the matrix elements is not equal to 1
	Speed
	Correction of overflow and underflow
Display pseudo-colour	Ability to change colour interactively
Generate series of file names automatically	
Stamp time-date on images	
Clear frame buffers	
Copy images among frame buffers	
Translocate images in frame buffers	
Compress/decompress image files	
Change image size	
Superimpose two images in different colours	
Display scale bars	

5.1 Qualitative analysis

Fluorescence images, especially those of fluorescent analogues, often show a very low contrast. It is not uncommon for an image to contain only pixel values between 0 and 30 in an eight-bit frame buffer, which allows possible grey values from 0 to 255. The simplest approach for contrast enhancement is to expand linearly the pixel values within a specified range to fill the entire grey scale. Although the enhancement does not provide any new information, it greatly facilitates the visualization of faint structures. Similar methods are already used widely in polarization and DIC optics. The enhancement should be implemented through look-up tables in the image processor, and ideally should allow the user to change the extent of enhancement interactively while viewing the image.

A second highly useful technique facilitates the detection of motion and differences among images. When a small number of images are involved, as in double immunofluorescence, the images are simply loaded into several frame buffers and the display is changed back and forth among the images, either automatically or manually. Most image processors allow three or four images to be compared in this fashion. For the analysis of motion, it generally

17

requires the successive display of more than five images in a time frame shorter than that required for retrieving images from the computer disk. A simple technique to achieve this is to divide the frame buffers into multiple domains and load different images at a reduced resolution into each domain. The display is zoomed into individual domains, such that only one image is displayed at a time, and is switched rapidly among different images. The division of a 512 × 480 frame buffer into four 256 × 240 domains quadruples the number of images that can be loaded without a serious sacrifice in resolution, and allows motion pictures to be constructed with 12–16 frames in most image processors. Ideally, the computer program should allow repeated displays of a sequence of images in either the forward or backward direction, at a speed that can be altered while viewing the motion picture.

Various techniques have been developed for the sharpening of images or the removal of noise (28, 29). The success of such techniques depends heavily on the nature of the image. For fluorescence images of a limited signal/noise ratio, the main challenge is to distinguish structural details from random noise.

In our experience, removal of noise with simple convolution techniques usually results in a significant loss of image resolution. We have developed a simple approach for noise removal based on the usually gradual transition in the fluorescence intensity around true structures, but sharp changes around the noise. At each pixel, calculations are performed using its intensity value and neighbouring intensity values to find a most likely value, e.g. by least squares fitting into a second-order equation. The fit values are then averaged with the original pixel value to obtain the final value for the pixel. This method results in a significant reduction in random noise with only a slight decrease in the 'sharpness' of the image.

Sharpening of images involves both an increase in the rate of transition of pixel values and the removal of out-of-focus signals. The latter has been done with dramatic success for the imaging of chromosomes, through calculations based on through-focusing image series (30). Since the amount of out-of-focus signals may be limited in well-spread cells, it is uncertain whether equally impressive results can be obtained with this approach for the study of cytoskeletons in cultured cell lines. The increase in the rate of pixel value transition is readily achieved using convolution techniques. The most serious problem with commonly used matrices is the significant increase in noise and the possible creation of artefacts. Limited success is achieved in the author's laboratory with matrices such as

$$
\begin{array}{rrrrr}
0 & -1 & -1 & -1 & 0 \\
-1 & 2 & 2 & 2 & -1 \\
-1 & 2 & 5 & 2 & -1 \\
-1 & 2 & 2 & 2 & -1 \\
0 & -1 & -1 & -1 & 0
\end{array}
$$

that suppress noise in local domains but amplify changes over longer distances.

5.2 Quantitative analysis

A great deal of information can often be obtained by the simple measurement of fluorescence intensities at specific structures. With images stored as arrays of pixels in a frame buffer, it is relatively easy to obtain intensity values at specific points or as profiles along specific lines. An equally useful program allows the integration of intensities within an enclosed area of any irregular shape. A pointing device, such as a graphics tablet or mouse, is commonly used to facilitate the definition of points and lines in the image. Due to the presence of noise in most fluorescence images, it is usually more useful to calculate an average over a local area, e.g. over 5×5 pixels, rather than using values of single pixels for point measurements and linear profiles.

Common sources of error for intensity measurements are the non-uniform detection of fluorescence across the field and the deviation from linearity of many video cameras. The former is caused primarily by the non-uniform throughput of the optical system and the uneven response of the detector, and can be corrected by dividing the image with the image of a uniformly fluorescent object, e.g. a uniformly dispersed fluorescent solution. The problem of non-linearity is associated with the limited dynamic range of many cameras, and can be alleviated by transforming the intensity values with a look-up table. It should also be noted that many experiments require only relative comparison of intensities, and can be performed simply by using a constant, central region of the field and by staying well within the dynamic range of the detector.

Equally useful is the ability to determine the relative position and dimension of specific structures. This also allows the calculation of the rate of any movement. A simple approach is to define two specific points on the screen with a pointing device, and to ask the computer to calculate the distance based on the number of pixels between the points, the aspect ratio of pixels (typically 5:4), and the magnification. Similar programs can be developed to calculate the surface area of any irregular shapes based on the number of pixels and the magnification.

Two sources of error should be noted in dimension measurements. First, some intensified cameras, notably ISIT cameras, introduce significant distortions, especially near the periphery of the field. The error affects primarily the measurement of dimensions that approach the width of the field. Methods have been developed for the correction of such distortions (31), and have been implemented in some image processors (e.g. Inovision). However, the extent of error may be insignificant for many applications, and it may be easier just to avoid the measurement of objects near the periphery of the field.

The second source of error affects primarily the measurements of small objects approaching the limit of resolution of the microscope, and often shows a strong dependence on the intensity of the object. Specifically, the apparent size is affected not only by the point spread function of the microscope, but also by the frequency response and the blooming of the video camera (21). Thus the measurement of diameters of less than 0.5 μm can become unreliable, even when the density of pixels allows a much better resolution. The limitation is less serious for the measurement of minuscule displacements provided the signal/noise ratio is high, since in this case the position can be determined based on the intensity profile.

Ratio imaging (32), developed primarily for the analysis of intracellular ion concentrations, can also yield valuable information on the distribution of fluorescent staining or analogues. For example, the true distribution of the concentration of fluorescent molecules can be obtained by dividing the fluorescence image with an image of control fluorescent molecules (with a different wavelength) that assume a uniform distribution in the same cell. Otherwise, the apparent fluorescence intensity can be easily affected by the distribution of organelles and the variation in the thickness in different regions of the cell (33). Ratio imaging can also be used to compare the distribution of different structural components or protein isoforms, and to identify sites where specific subpopulations of molecules are localized. One recent example involves the microinjection of living cells with a fluorescent analogue, followed by the extraction of cells after a short period of incubation (34). The preferred sites of incorporation are mapped by dividing the image of injected analogues with the image of staining, which indicates the overall distribution of the corresponding endogenous component. A high ratio thus represents a high incorporation rate on a per unit mass basis. Without the ratio imaging, it would be impossible to determine the true active sites of incorporation, since the apparent intensity of the analogue is determined by the rate of incorporation as well as the distribution of structures, both of which can vary dramatically in different regions of the cell.

Many technical aspects of ratio imaging have been discussed in a recent article by Bright *et al.* (32). An important problem which can seriously affect the performance of ratio imaging is the misregistration of images. This problem is most serious when the two images are recorded with different dichroic mirrors and barrier filters, since slight inconsistencies in the alignment of these components can cause significant shifts in images. Sophisticated algorithms have been developed to align two images (31), but the implementation of such programs in image-processing systems is not a trivial task.

We have developed a simple strategy for aligning images that relies on visual judgment. Briefly, the two images for ratioing are displayed back and forth at a high speed. Misregistration is readily visualized as wobbling of the images, and can be eliminated by moving the images relative to each other until any wobbling disappears. This method effectively removes misregistrations due to translation, but not those caused by rotation, differential magnification, or distortion. However, these latter problems are usually insignificant over limited domains of the cell.

6. Conclusion and prospects

Although immunofluorescence is a well-developed technique, there is still great potential for the development of new fluorescent ligands or peptides as agents for the detection of specific components (13). In addition, many potential applications of fluorescent analogue cytochemistry remain to be explored. For example, cells microinjected with fluorescent analogues can be readily treated with various chemical agents or probed with other micro-manipulation methods for mechanical or electrophysiological properties. The fluorescence signals can be analysed with photobleaching or photoactivation and various spectroscopic techniques, and provide information much beyond the resolution of the light microscope, e.g. on the conformation, association, and mobility of molecules. The method is also easily combined with ion imaging for the analysis of possible roles of second messengers in cytoskeletal dynamics. The use of caged second messengers represents a particularly attractive approach for such studies.

Acknowledgements

The author wishes to thank Dr Douglas Fishkind and Mr Mitchell Sanders for helpful comments. The research is supported by NIH grants GM32476, GM41681, and a grant from the Muscular Dystrophy Association.

References

1. Wang, K., Feramisco, J., and Ash, J. (1982). *Meth. Enzymol.*, **85**, 514.
2. Osborn, M. and Weber, K. (1982). *Meth. Cell Biol.*, **24**, 97.
3. Wang, Y.-L., Heiple, J. M. and Taylor, D. L. (1982). *Meth. Cell Biol.*, **25**, 1.
4. Simon, J. R. and Taylor, D. L. (1986). *Meth. Enzymol.*, **134**, 487.
5. Kreis, T. E. (1986). *Meth. Enzymol.*, **134**, 507.
6. Wadsworth, P. and Salmon, E. D. (1986). *Meth. Enzymol.*, **134**, 519.
7. Wang, Y.-L. (1989). *Meth. Cell Biol.*, **29**, 1.
8. Small, J. V., Rinnerthaler, G., and Hinssen, H. (1982). *Cold Spring Harbor Symp. Quant. Biol.*, **46**, 599.
9. Schliwa, M. and Blerkom, J. van (1981). *J. Cell Biol.*, **90**, 222.
10. Gilow, H. and Sedat, J. W. (1982). *Science*, **217**, 1252.
11. Hegman, T. E., Lin, J. L.-C., and Lin, J. J.-C. (1988). *J. Cell Biol.*, **106**, 385.
12. Cooper, J. A., Loftus, D. J., Frieden, C., Bryan, J., and Elson, E. L. (1988). *J. Cell Biol.*, **106**, 1229.

13. Wang, Y.-L. and Sanders, M. C. (1990). In *Noninvasive techniques in cell biology*, (ed. K. Foskett and S. Grinstein), p. 117. Wiley-Liss, New York.
14. Means, G. E. and Feeney, R. T. (1971). *Chemical modification of proteins.* Holden-Day, San Francisco.
15. Wang, Y.-L. and Taylor, D. L. (1980). *J. Histochem. Cytochem.*, **28**, 1198.
16. Lamb, N. J. C., Fernandez, A., Conti, M. A., Adelstein, R., Glass, D. B., Welch, W. J. and Feramisco, J. R. (1988). *J. Cell Biol.*, **106**, 1955.
17. McKenna, N. M. and Wang, Y.-L. (1989). *Meth. Cell Biol.*, **29**, 195.
18. Aubin, J. E. (1979). *J. Histochem. Cytochem.*, **27**, 36.
19. Graessman, A., Graessman, M., and Mueller, C. (1980). *Meth. Enzymol.*, **65**, 816.
20. Lee, G. M. (1989). *J. Cell Sci.*, **94**, 443.
21. Inoue, S. (1986). *Video microscopy.* Plenum, New York.
22. Taylor, D. L. and Salmon, E. D. (1989). *Meth. Cell Biol.*, **29**, 208.
23. Wampler, J. E. and Kurz, K. (1989). *Meth. Cell Biol.*, **29**, 239.
24. Spring, K. R. and Lowy, R. J. (1989). *Meth. Cell Biol.*, **29**, 270.
25. Aikens, R. S., Agard, D. A., and Sedat, J. W. (1989). *Meth. Cell Biol.*, **29**, 292.
26. Tsay, T.-T., Inman, R., Wray, B. E., Herman, B., and Jacobson, K. (1990). In *Optical microscopy for biology*, (ed. B. Herman and J. Jacobson), p. 219. Wiley-Liss, New York.
27. Cao, L.-G. and Wang, Y.-L. (1990). *J. Cell Biol.*, **111**, 1905.
28. Gonzalez, R. C. and Wintz, P. (1977). *Digital image processing.* Addison-Wesley, Reading, MA.
29. Pratt, W. K. (1978). *Digital image processing.* Wiley, New York.
30. Agard, D. A., Hiraoka, Y., Shaw, P., and Sedet, J. W. (1989). *Meth. Cell Biol.*, **30**, 353.
31. Jericevic, Z., Wiese, B., Bryan, B., and Smith, L. C. (1989). *Meth. Cell Biol.*, **30**, 48.
32. Bright, G. R., Fisher, G. W., Rogowska, J., and Taylor, D. L. (1989). *Meth. Cell Biol.*, **30**, 157.
33. DeBiasio, R. L., Wang, L.-L., Fisher, G. W., and Taylor, D. L. (1988). *J. Cell Biol.*, **107**, 2631.
34. Cao, L.-G. and Wang, Y.-L. (1990). *J. Cell Biol.*, **110**, 1089.
35. Wang, Y.-L. (1991). *Meth. Enzymol.*, **196**, 497.

2

An ultrastructural approach to understanding the cytoskeleton

JOHN H. HARTWIG

1. Introduction

The shape and movement of cells is directed by a cytoplasmic scaffolding of fibres which has become known as its cytoskeleton. It includes the nuclear matrix and its DNA, all major filament systems of cells—microfilaments, microtubules, and intermediate filaments—and the many proteins which are associated with and regulate these three fibre systems. The cytoskeleton also includes receptors that are coupled, or can become coupled with occupancy, to underlying cytoskeletal fibres. All components are assembled into a three-dimensional structure which defines the cell's architecture and surface topology.

From a practical point of view, the cytoskeleton has been defined as all components that are insoluble when a cell is delipidated with detergents in buffers that mimic cytoplasmic ionic conditions. Considerable information is now available on the biochemistry, structure, and genetics of many of the molecular components of the cytoskeleton. The morphology of cytoskeletal proteins has also been addressed in many studies and a variety of methodology developed which preserves three-dimensional cytoskeletal structure. Molecular cloning techniques have rapidly increased our knowledge of the primary structure of microfilament, intermediate filament, and microtubule accessory proteins. Moreover, recombinant DNA and bio-chemical advances have permitted the isolation of these proteins and led to the generation of antibodies reactive toward these immunogenic epitopes. These antibody reagents are now available to locate individual regulatory proteins and allow their biochemical functions to be related to their cytoskeletal location.

23

2. Preparation of cytoskeletal fractions

2.1 Detergent solubilization or permeabilization of membranes

Detergent treatment, first used by Brown *et al.* (1), is the approach of choice for separating cytoskeletal elements from membrane and soluble cellular proteins. When cells are attached to a plastic or glass surface, detergent treatment leaves adherent, insoluble residues that are easily processed for the light or electron microscope. Permeabilization of cells in suspension with a view to observing them under the electron microscope is also possible, but the resultant cytoskeletons must subsequently be attached to a support surface for further processing. The type of detergent to be employed, as well as the amount used, depends on the desired amount of membrane to be removed. For a further discussion of membrane solubilization, see Chapter 6 of this volume. Cells are permeabilized in buffers thought to mimic intracellular osmolarity. Reagents which stabilize cytoskeletal components and/or inhibit proteases should be included. Examples of reagents that stabilize fibre systems are phalloidin (or phallacidin) and taxol. Phalloidin binds stoichiometrically to actin subunits in filaments, stabilizing them against conditions that would normally depolymerize them, in particular, upon dilution below their critical concentration (2, 3, 4), which occurs during detergent treatments involving a large volume of solution bathing a relatively small number of cells. Taxol, in a similar fashion, stabilizes microtubules (5). Optimal stabilization of cytoskeletal structure occurs in the presence of concentrations of EGTA sufficient to chelate calcium released during delipidation. The integrity of the actin filament networks at the cell cortex is exceedingly sensitive to micromolar levels of calcium because of calcium-activated microfilament-severing proteins and calcium-inactivated filament cross-linking proteins in cells (6). Other components in the extraction buffer may be critical in preserving associations of proteins with cytoskeletal elements. These include nucleotides, reducing compounds, phosphatase inhibitors, and others. The polypeptide composition of detergent soluble and insoluble phases can be analysed by SDS-PAGE (7) and the distribution of a protein between the phases quantitated by immunoblotting from these gels (8). The effect of buffer components on the extractability of a protein, therefore, can be directly determined.

Many buffers have been used in combination with Triton X-100, or other non-ionic detergents such as NP-40, to permeabilize cells. All contain calcium-chelating reagents. Some examples include:

- 137 NaCl, 5 mM KCl, 1.1 mM Na_2HPO_4, 0.4 mM KH_2PO_4, 4 mM $NaHCO_3$, 5.5 mM glucose, 2 mM $MgCl_2$, 2 mM EGTA, 5 mM Pipes, pH 6.1 (9);
- 100 mM NaCl, 300 mM sucrose, 10 mM Pipes, 3 mM $MgCl_2$, 1 mM EGTA, 4 mM vanadyl riboside complex, 1.2 mM PMSF, pH 6.8 (10);

- 100 mM Pipes, 0.5 mM $MgCl_2$, 0.1 mM EDTA, 4 M glycerol, pH 6.9 (11);
- 60 mM Pipes, 25 mM Hepes, 2 mM $MgCl_2$, and 10 mM EGTA, pH 6.9 (12).

To guarantee reproducible cytoskeletal preparations, detergents are generally applied in generous quantities, removing the maximum amount of membrane lipid. Therefore, as mentioned above, it is the composition of the extraction buffer that is critical in determining the ultimate insolubility of proteins in the cytoskeleton.

2.1.1 Solubilization of membranes with Triton X-100

For a standard cytoskeleton preparation protocol, we have selected the buffer of Schliwa *et al.* (12), also adding 5 μM phallacidin, and the protease inhibitors 4.2 nM leupeptin, 1 mM benzamidine, and 12.3 nM aprotinin (PHEM buffer). This buffer preserves all actin which was filamentous prior to extraction. A 10× stock is prepared, and after sterile filtration has a shelf life of months. Glutaraldehyde (0.1–1 per cent) or formaldehyde can be added to the extraction solution if desired.

The critical micelle concentration for Triton is 240 μM. Cells adherent to a surface are most easily processed. The type of surface to which the cells are attached depends on the subsequent processing techniques. For light microscopy, 10 mm round or 22 mm square glass cover slips are used. For rapidly-frozen, freeze-dried, metal-coated cytoskeletons, small, 3–5 mm diameter glass cover slips are best. For negatively-stained cytoskeletons, cells are first allowed to attach and spread on nickel electron microscopy grids that have been coated with either a formvar film (the film is prepared from 0.3 per cent solution of formvar in ethylene dichloride) or a nitrocellulose (parlodion or collodion) film (a drop of 2.5 per cent nitrocellulose in amyl acetate spread on the surface of water) then permeabilized with detergent. To attach cytoskeletons from suspensions of cells an adherent surface must be used. Glass cover slips are first made hydrophilic by glow discharge, then freshly coated with a 1 mg/ml polylysine hydrobromide (> 70 000 M_r) solution. Cytoskeletons can then be attached with minimal damage by placing a drop of solution containing the cytoskeletons on the grid or glass surface and allowing them to settle gently. Centrifugation of cytoskeletons can also be used to cause them to adhere to glass. Cover slips with or without grids attached to them are covered with buffer containing cytoskeletons and centrifuged in the bottom of wells of a 96-well plastic culture plate. This yields a uniform coat of cytoskeletons on the cover slip. See *Protocol 1*.

Protocol 1. Preparation of cytoskeletons of adherent cells
(All procedures are carried out at 37 °C)

  1. Using double-sided adhesive tape, attach approximately one third of an acid-cleaned and glow-discharged cover slip to the bottom of a 10 cm diameter Petri dish.

Protocol 1. *Continued*

2. Place a drop of cell suspension on each cover slip, incubate for 1–2 min, and wash away non-adherent cells.

3. Allow the cells to spread and move, or expose to agonists or inhibitors.

4. Wash cells in Petri dish rapidly with PHEM buffer. (Washing is carried out by evacuating the liquid with a Pasteur pipette, then immediately replacing it with new solutions.)

5. Permeabilize cells with PHEM buffer containing 0.75 per cent Triton X-100, protease inhibitors, and phallacidin for 2 min.

6. Wash cells twice with PHEM buffer.

7. Fix the cells for 10 min in PHEM buffer containing 1 per cent glutaraldehyde.

8. Wash the cytoskeletons with distilled water containing 0.1 μM phallacidin for negative staining or rapid freezing (see Section 3.1).

9. Quench unreacted aldehydes with a 2 min rinse in PHEM buffer containing 1 mg/ml sodium borohydride.

10. Wash cells three times in 0.154 M sodium chloride, 10 mM sodium phosphate buffer (pH 7.4) containing 1 mg/ml of IgG-free bovine serum albumin (PBS–BSA).

The cytoskeletons are now ready for incubation with specific antibodies (Section 5).

Cytoskeletons prepared with Triton X-100 have the following advantages and disadvantages:

- *Advantages*: Rapid, maximal lipid removal is achieved; cytoskeletal-based antigens are readily accessible to antibodies and immunogold probes; biochemical characterization of the composition of cytoskeletons is straightforward by SDS-PAGE (7) of adherent and soluble detergent fractions.

- *Disadvantages*: The major limitation is that cytoskeleton–membrane interactions may be disrupted by the detergent.

2.1.2 Saponin treatment

Saponin, a detergent semi-selective for cholesterol (13, 14), is used to permeabilize the plasma membranes of cells without completely solubilizing the lipid bilayer. Surprisingly, these lightly permeabilized cells retain many biological functions that can now be studied in greater detail by adding inhibitors previously impermeant to intact cells. The degree of permeability of the membrane of saponin-treated cells varies, depending on the concentration used and the time of treatment. Membrane permeability sufficient for entry of IgG or myosin subfragment 1 and for leakage of cytoplasmic actin monomers from cells is achieved with buffers containing 0.01–0.02 per cent saponin for 5–10 min (15, 16). Similarly, low doses or short Triton X-100 exposure times can be used to remove membrane selectively.

Saponin-treated cells have the following advantages and disadvantages:

- *Advantages*: More membrane–cytoskeletal interactions are preserved than with non-ionic detergent (15). Correlation of structural analysis with motility in model systems, in particular, by adding reagents to block processes inaccessible in the intact cell, is now feasible.

- *Disadvantages*: Cytoskeletons must be fractured to visualize interiors. Technically, obtaining the desired fracture line is a rare and random event on adherent cytoskeletons.

2.1.3 Wet cleavage of cells

The mechanical disruption of cells has proved useful for studying the interaction of cytoskeletal elements and the cytoplasmic side of the plasma membrane (*Figure 1*) as well as for viewing individual proteins associated with the cytoplasmic side of the membrane (e.g. the H^+-ATPase complex in the intercalated cells of the kidney collecting duct and clathrin coats on membrane (17, 18). Membrane–cytoskeletal fragments from adherent cells have been prepared by shearing them with liquid streams (19) and by using mechanical shear, such as scraping their apical surface with the edge of a glass slide (17) or attaching a cover slip to the exposed surface and removing it (unroofing, 20). See *Protocol 2*.

Protocol 2. Method for wet cleavage of cells between two cover slips

1. Using double-sided adhesive tape, attach approximately one third of an acid-cleaned and glow-discharged cover slip, cell side up, to the bottom of a 10 cm diameter Petri dish.

2. Wash cells into PHEM buffer.

3. Place a second cover slip previously coated with polylysine on top of the first one.

4. Squeeze the two cover slips together without sliding them relative to one another, using moderate pressure applied at the middle of the top cover slip with the tips of a pair of tweezers.

5. Remove the polylysine-coated cover slip, which will be decorated with fragments from the apical membrane of cells. The original cover slip holds the remains of the broken cells, i.e. the adherent cells and intact cells not cleaved during this process.

6. Wash the membrane-cytoskeletal fragments for 2 min in PHEM buffer, fix, decorate with antibody, and prepare for electron microscopy.

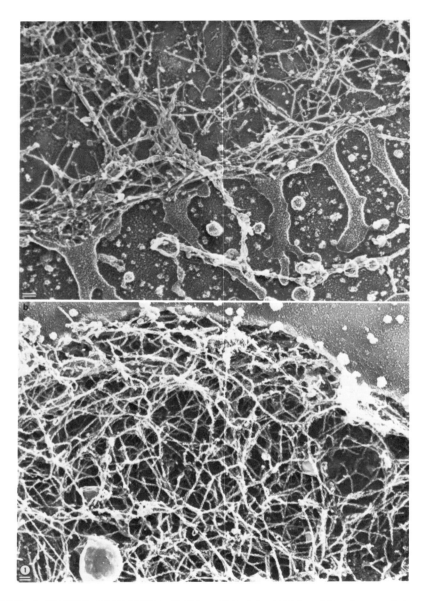

Figure 1. Electron micrographs showing the cytoplasmic side of a wet-cleaved (unroofed) polymorphonuclear leukocyte. Polymorphonuclear leukocytes, adherent to glass, were unroofed by attaching and removing polylysine-coated cover slips. (a) This representative electron micrograph shows regions of membrane containing actin filaments and bare regions of membrane. (b) Demonstration (by myosin S1 labelling) that fibres interacting with the plasma membrane are actin. The bar is 0.1 μm

The preparation described in *Protocol 2* has the following advantages and disadvantages:

- *Advantages*: Direct access to and visualization of cytoskeleton–cytoplasmic side of plasma membrane interface is afforded. Comparison of the interaction of filaments with apical and basal cell surfaces is possible.
- *Disadvantages*: Unroofing is random, and many cells are not disrupted. Therefore, biochemical analyses are not possible. This procedure can only be used on adherent cells or the apical membrane of epithelial tissue cells.

3. Preparation of cytoskeletons and cytoskeletal fractions for the electron microscope

Whole mounts of cells can be viewed using heavy metal stains or after metal coating. If cells are to be metal-coated, the water must be removed by either critical-point drying or rapid freezing. Of the two, the latter provides better results; see Section 3.3.

3.1 Negative staining

Optimal negative staining of cytoskeletons (*Figure 2a*) can be obtained by following the staining protocol of Small (9). Three stains have been used with equivalent results: 1 per cent uranyl acetate, 2 per cent sodium silicotungstate, or 2 per cent phosphotungstic acid. *Protocol 3* describes the negative staining procedure.

Protocol 3. Negative staining of cytoskeleton preparations

1. Hold the grid with the tips of clamped forceps.
2. Incubate the grid in distilled water containing 5 μM phallacidin and rinse three times with the same solution.
3. Wash the grid twice with a few drops of 40 μg/ml bacitracin.
4. Wash the grid with the negative staining solution by adding a few drops of stain to the grid.
5. After the final drop, remove the staining solution by touching the grid–forceps junction with filter paper.
6. View grid at 80 kV. Care must be taken when viewing negatively stained cells. The cell is first illuminated at low magnification, uniformly warming it and the grid square. When increasing the magnification, the beam is placed directly in the middle of the cytoskeleton, spread, and then slowly moved to the periphery of the cytoskeleton.

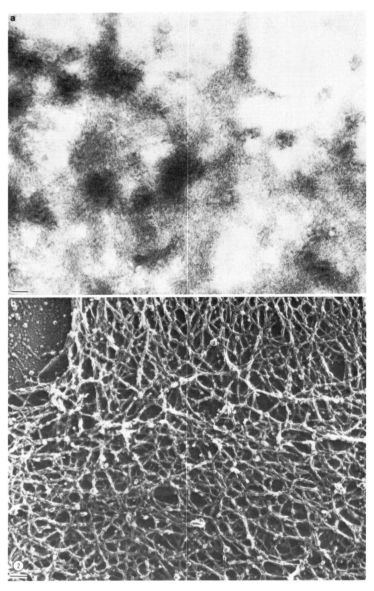

Figure 2. Comparison of the organization of actin in the periphery of macrophage cytoskeletons after negative staining (a) or rapid freezing, freeze-drying, and rotary metal coating with platinum and carbon (b). The plasma membrane was dissolved with 0.75 per cent Triton X-100 in a physiological buffer containing 10 mM EGTA. The magnifications of the two micrographs are identical. The bar is 0.1 μm. (a) Organization of actin filaments at the cytoskeletal periphery after negative staining with 2 per cent sodium silicotungstic acid. (b) Organization of actin filaments at the macrophage periphery in a platinum replica. The cytoskeleton was then prepared for electron microscopy by brief fixation, rapid-freezing, freeze-drying at −80 °C and coating with platinum and carbon. Two types of filament ultrastructure are shown. The bulk of the actin filaments at the periphery of the cell organize in an orthogonal network of small porosity. Filopods at the edge of the cell contain filaments aligned in parallel.

30

3.2 Rapid freezing, freeze-drying

Cover slips containing cells or cytoskeletons are washed with distilled water containing 1 μM phallacidin and the cover slips mounted on aluminium freezing tabs numbered on their bottom side with a suitable pen. To avoid crushing the specimen, a cushion is placed with attached filter paper between the aluminium tab and the cover slip: for example, 0.4 mm thick sections of fixed rat lung or slices of 1–2 per cent agarose have been used. Specimens are then frozen by slamming them on to a liquid helium-cooled copper block (11). A number of alternative types of freezing apparatus are now available, all operating by a similar approach. Once frozen, samples can be stored for an indefinite time in liquid nitrogen. The optimal storage system uses 35 mm tin film canisters. The removal screw top is punctured and a long string attached through the hole. This string is used to retrieve the canisters from a liquid nitrogen Dewar. The sides of the canister are punctured in two places, each about 2.5 cm from the bottom of the canister (so that it does not float on the liquid nitrogen). Freezing tabs containing the frozen cover slips are transferred to a liquid nitrogen-cooled stage of a freeze fracture apparatus. The stage temperature is raised from liquid N_2 temperature to −80 °C to −100 °C for 20–45 min, and then the cover slips are rotary coated with platinum or tantalum–tungsten from a source mounted at an angle (relative to the sample) of 25–45 ° (depending on the amount of edge contrast desired) and with carbon at 90 ° (*Figure 2b*).

3.2.1 Choice of metal coating

i. Platinum

Platinum is the most widely used metal for coating dried specimens (*Figure 3a*) (21, 22). Its advantage is ease of use, high contrast, and stability. Its major disadvantage relative to tantalum–tungsten (below) is a heterogenous particle size of between 0.5 and 2 nm when deposited onto the sample. Since the resolution is defined by the large particles, the resolution of platinum replicas is limited to about 2 nm. The grain size of platinum on a sample is, however, temperature dependent, with grain size becoming finer and more uniform as the temperature at which the sample maintained is lowered (−160 to −190 °C). The high contrast of platinum replicas can also be problematic when locating antigens with colloidal gold probes. Gold particles > 8 nm are visible in thin replicas (1.0–1.5 nm), but gold particles of smaller sizes are difficult to resolve.

ii. Tantalum–tungsten alloy

Replicas of tantalum–tungsten are composed of fine, homogeneous grains (*Figure 3b*) (23). Since this alloy is not as dense as platinum, replicas of identical thickness have lower contrast than platinum, although thicker

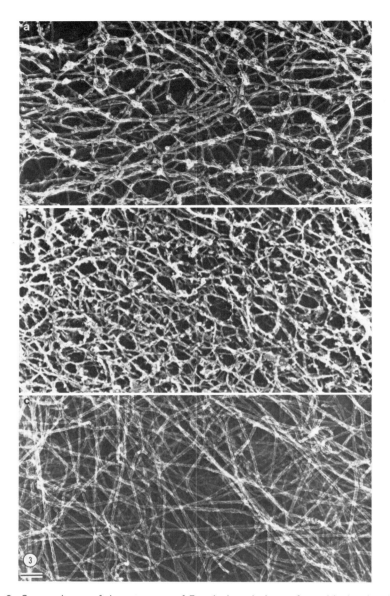

Figure 3. Comparisons of the structure of F-actin in solutions after critical-point drying and freeze drying. (a), (b) 3 µl of 24 µM F-actin was polymerized with 2 mM MgCl$_2$. It was rapidly frozen without fixation, the frozen sample cleaved with a liquid nitrogen-cooled knife (Delaware Diamond) and freeze-dried for 18 min at −98 °C. The sample was rotary-coated with either 1.5 nm of platinum (a) or 1.2 nm of tantalum−tungsten alloy (b) and 4 nm of carbon without rotation. Note the high resolution of the actin polymers. These double-helical filaments complete a twist every 75 nm or every 28 actin subunits. This twist, as well as the subunit structure of the filaments, can be visualized. (c) 24 µM F-actin prepared for the electron microscope by critical-point drying.

32

replicas, of 1.5–2.0 nm, can be used to enhance contrast. Tantalum–tungsten replicas 0.1–1.4 nm in thickness provide maximal resolution and are ideally suited for immunogold labelling studies, maximizing the contrast between tantalum–tungsten-coated structures and colloidal gold particles, even particles of 3–5 nm diameter. The uniform and small grain size of this alloy also increases the depth the metal penetrates into the interstices of three-dimensional fibre networks, stabilizing them to a greater extent. Tantalum–tungsten replicas are, however, less stable than platinum replicas and care must be taken when cleaning them. Common bleach will dissolve tantalum–tungsten alloy replicas in a few seconds. This alloy is also susceptible to hydrofluoric acid, albeit slowly. Dissolution by hydrofluoric acid therefore eliminates this metal alloy as a coating material for molecules using the mica flake technique (24, 25). Replicas containing mica fragments must sit in this acid overnight to dissolve the mica. Removal of tissue from under tantalum–tungsten replicas can be accomplished using chromic acid.

iii. Carbon

Carbon is always applied immediately after the primary metal. Although carbon is considerably less dense than platinum or tantalum–tungsten alloy (2.2 g/ml as opposed to 22.9 or 17.7 g/ml), care should be exercised to apply a thin film of carbon. The best results are obtained when a carbon thickness of 2.5–3.5 nm is applied. If the sample is rotary-shadowed first with platinum or tantalum–tungsten, then unidirectional carbon film applied to 90° provides the best resolution.

3.2.2 Cleaning of replicas

After drying and application of metal to the cytoskeletons, they remain attached to the glass surface. To separate metal replicas from the underlying glass surface, the cover slip–cytoskeleton–replica sandwich is carefully floated on to a 25 per cent hydrofluoric acid solution in a small plastic Petri dish. Once the replica separates from the cover slip, it is moved into distilled water. Replica transfer is accomplished using Pasteur pipettes whose ends are flame-sealed and bent at 90° about 1 cm from the end. If the replica is tantalum–tungsten (sensitive to hydrofluoric acid), its separation from the cover slip is facilitated by carefully grasping the edge of the cover slip with tweezers and slowly tilting the cover slip into the liquid as the replica detaches. Replicas are passed into other cleaning solutions by slowly bringing the end of the Pasteur pipette under them, picking them up, and carefully placing the pipette tip with the replicas in a new solution. After cleaning with bleach or chromic-sulphuric acid, replicas are passed through at least three washes of distilled water. To minimize disruption of the replica caused by differences in surface tension between different aqueous solutions, all solutions are first touched with the tip of a platinum wire previously dipped in Photoflo (Kodak).

Cleaned replicas are picked up on copper grids. Maximal resolution, but minimal stability, is obtained using uncoated grids which are first cleaned in a 10 per cent formic acid solution to remove the oxidized layer. Maximal stability of the replica is obtained, however, using grids coated with either a 0.25 per cent formvar film or nitrocellulose film. The grids are then coated with a thin carbon film and subjected to glow discharge to make their surface hydrophilic.

3.3 Critical-point drying

Critical-point drying has been used to dehydrate cytoskeletons for electron microscopy (26). Although the technique is widely used, questions remain about artefactual shrinkage and collapse during critical-point drying despite efforts to eliminate water contamination in the liquid carbon dioxide drying solution. Clearly, the diameters of cytoskeletal elements prepared by this approach are highly variable when compared to those in freeze-dried samples (*Figure 3c*) (27). See *Protocol 4*. This approach has been used on both intact, fixed cells (29) and fixed cytoskeletal preparations.

Protocol 4. Critical-point drying

1. Prepare cytoskeletons from cells attached to formvar-coated grids.
2. Dehydrate samples using increasing concentrations of ethanol: 10 per cent, 25 per cent, 50 per cent, 75 per cent, 90 per cent, 100 per cent.
3. Wash three times in 100 per cent ethanol.
4. Place the sample in the critical-point drying apparatus and purge the ethanol with a 5–10 min wash of liquid carbon dioxide.
5. Discontinue the carbon dioxide wash and dry the sample under pressure by increasing the temperature of the liquid.
6. View the dried samples directly in the electron microscope or transfer them to a freeze-fracture machine and coat wth platinum or tantalum–tungsten. Metal coating improves the uniformity of the fibres, so electron beam damage cannot be disregarded.

3.4 Dry cleaving of cytoskeletons

An interesting modification for cleaving cells has been developed by Mesland and Spiele (30). Intact, fixed cells were critical-point dried and then decapitated by attaching a piece of tape to their exposed membrane surface and removing it. Theoretically, this approach could be used on saponin- and antibody-treated cells and on cytoskeletons which have been frozen and freeze-dried. Double-sided adhesive tape on the knife holder in a freeze-fracture apparatus could be lowered onto the sample to effect cleavage. The sample could then be coated with metal.

3.5 Cytoskeletal structure in intact, unfixed cells

Cells or pieces of tissue are frozen (cryofixed). They are then transferred to the stage of a freeze-fracture machine, scraped with a liquid nitrogen-cooled knife, and cytoskeletal elements revealed by freeze-drying and metal-coating (31, 32). There are two limitations to this approach. First, the amount of freeze-drying is self-limiting as the intracellular salts and soluble proteins begin to accumulate and obscure cytoskeletal elements. The depth of useful etch can be increased by brief treatment of cells or tissue with distilled water just prior to cryofixation (*Figure 4*). Such treatment, however, may alter cell structure. Second, once the sample has been frozen, immunogold labelling must be done by either freeze substitution (*Protocol 5*) or controlled freeze-drying (molecular distillation); see Section 3.5.1.

Protocol 5. Freeze-substitution

1. At liquid nitrogen temperature, place cryofixed specimens frozen side down in a scintillation vial in a solution of 10 per cent acrolein in acetone containing 0.2 per cent tannic acid. Cover with liquid nitrogen.

2. Place the scintillation vial in a styrofoam container, cover it with liquid nitrogen, and leave it overnight (10–24 h) in a freezer at −80 °C.

3. Warm the sample to −55 °C, wash with acetone at this temperature and place in 0.2 per cent osmium tetroxide in acetone; then warm to −20 °C.

4. Transfer the sample to glutaraldehyde (1–10 per cent) in 50 per cent acetone/40 per cent methanol at 0 °C.

5. Perform dehydration and embedding steps as described in Section 4 and *Protocol 6*.

3.5.1 Molecular distillation

Cryofrozen cells are slowly brought to room temperature *in vacuo* (34, 35) in a specialized drying apparatus that controls both vacuum and sample temperature. Once dehydrated and at room temperature, they can be fixed with either osmium or paraformaldehyde vapours and then infiltrated with resin without dehydration. Molecular distillation can be purchased as a service from Life Cell Corporation. This technique appears to provide a high degree of membrane preservation and antigenicity. Advantages of this technique include rapid fixation (cryofreezing) and a lack of exposure to organic solvents.

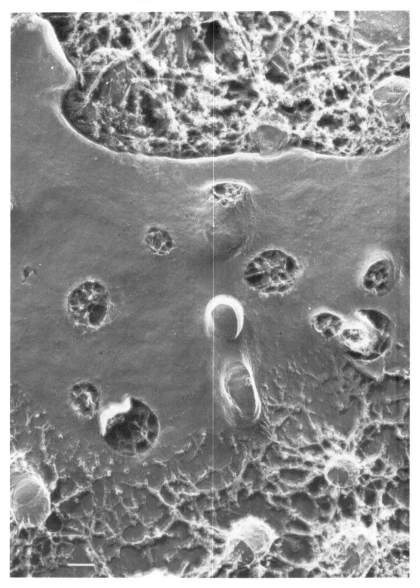

Figure 4. The organization of cytoplasmic fibres which support membrane microvilli and folds of membrane on the surface of granular cells of the toad bladder mucosal epithelium. A piece of living bladder was placed on a rapid-freezing tab, rapidly frozen, transferred to a stage cooled with liquid nitrogen, warmed to −120 °C, fractured with a diamond knife, and freeze-dried at −98 °C for 18 min. The sample was rotary coated with 1.5 nm of platinum at 45 ° and 10 nm of carbon at 90 °. The fracture line passed into the membrane, cutting off the tops of microvilli. Note the cytoskeletal system supporting the microvilli. The bar is 0.1 μm

36

4. Preparation of cell sections

4.1 Infiltration with Lowicryl K4M

Cells to be embedded for electron microscopy are fixed for 10 min in 1 per cent glutaraldehyde in 10 mM phosphate buffer, pH 7.4, and embedded in Lowicryl K4M (36, 37) using a rapid polymerization protocol. Other fixatives that have been shown to give good antigenicity include:

- 1 or 2 per cent glutaraldehyde in 0.1 M sodium cacodylate, pH 7.0;
- PLP fixative:
    150 ml of 5 mM sodium phosphate buffer, pH 7.4
    2.2 gm lysine
    428 mg sodium periodate
    50 ml 8 per cent paraformaldehyde.

For dehydration, wash the sample six times, as follows:
1. buffer three times, 5 min each;
2. 50 per cent dimethylformamide (glass only) or ethanol (plastic), 10 min;
3. 75 per cent dimethylformamide (glass only) or ethanol (plastic), 10 min;
4. 90 per cent dimethylformamide (glass only) or ethanol (plastic), 10 min.
Embedding in Lowicryl K4M is carried out as in *Protocol 6*.

Protocol 6. Embedding in Lowicryl K4M

1. Prepare K4M Lowicryl Kit (Polysciences or Biorad) by mixing 2 g of Part A with 13 g of Part B and adding 0.75 g Activator.
2. Wash samples in K4M:DMF (1 volume of K4M to 2 volumes of DMF or ethanol) for 10 min.
3. Wash samples in K4M:DMF (1 volume of K4M to 1 volume of DMF or ethanol) for 15 min.
4. Wash samples in 100 per cent K4M twice, for 25 min each time.
5. Transfer pellets to gelatin capsules for polymerization in fresh K4M. Suspend the capsules on small wire loops on a metal test-tube rack.
6. Polymerize for 45 min at room temperature using a 15 W UV light emitting at 360 nm (General Electric Blacklite or Philips 15 W TLD 15/05). Rotate the capsules 180 °C after 25 min.

To label the surface of thin sections with gold, the sections are placed on the surface of 10 µl drops of 10 µg/ml of anti-protein IgG in PBS (affinity-purified IgG if possible) containing 1 per cent BSA, pH 8.2 (PBS–BSA), and incubated for 1 h at 25 °C. Drops are maintained in a humidified, sealed petri dish on a parafilm surface. Unbound IgG is removed by washing the sections three times by transferring them to drops of PBS–BSA for 5 min each. They are then transferred to 10 µl drops of PBS–BSA containing gold particles coated with either anti-primary antibody-IgG or protein A-coated gold. Following the incubation with immunogold, the sections are washed twice in PBS and three times in distilled water, then stained with 2 per cent uranyl acetate for 7 min, followed by 1 per cent lead citrate for 1 min.

4.2 Resinless sections

Techniques have been described for infiltrating specimens in resins that become sufficiently hard to be sectioned but from which the specimens can be removed after sectioning. Wolosewick and De Mey (38) infiltrated cells with polyethylene glycol (PEG-4000) and hardened it with a 3 h incubation at 55 °C. The PEG resin was removed from the thin section using ethanol and the residue critical-point dried. PEG sections have proved to be difficult to work with and, more recently, Penman and coworkers (39, 40, 41) have used diethylene glycol distearate (DGD) to infiltrate cytoskeletal preparations in a similar fashion. After sectioning, DGD is removed with butanol. Sections are critical-point dried and viewed in the electron microscope after carbon or platinum–carbon coating. Sections can also be prepared for the electron microscope with heavy metal stains and can be rehydrated after removal of the resin and treated with antibodies and immunogold probes. In the method described in *Protocol 7*, cytoskeletons are prepared with detergents and fixed with glutaraldehyde or paraformaldehyde.

Protocol 7. Resinless sections: embedding in DGD

1. Dehydrate using increasing ethanol concentrations (see *Protocol 4*), with three final washes of 100 per cent ethanol.
2. Remove ethanol and gradually replace with butanol by washing sequentially in solutions of butanol/ethanol (1:2, v/v), then butanol/ethanol (1:1, v/v), followed by washes in 100 per cent butanol; 15 min each.
3. Infiltrate gradually with DGD at 60 °C using a wash series of butanol/ DGD (2:1, v/v), then butanol/DGD (1:2, v/v), then three washes in DGD; 1 h each.
4. Allow the resin to solidify and cut sections from it.
5. Pick up the sections on grids coated with carbon-nitrocellulose, and remove DGD resin using 100 per cent butanol.
6. Process sections for electron microscopy by either critical-point drying (see *Protocol 4*) or rapid freezing.

John H. Hartwig

5. Immunogold labelling

Colloidal gold particles are the label of choice for the electron microscope. They are electron dense, different sizes of particles can be easily prepared (2–20 nm), and they can be coated with a variety of protein reagents. Although it is possible to coat gold particles directly with IgG and Fab' moieties, labelling is inefficient because of limited particle mobility. Consequently, antigens are first amplified with specific antibodies and detected using a sandwich technique employing gold particle-coated IgGs reactive against the primary IgG (42, 43) or proteins such as *Staphlococcus aureus* protein A (37) or protein G, which bind IgG. Avidin- or anti-biotin IgG-coated gold particles can also be used to visualize biotinylated IgGs or other biotinylated molecules.

5.1 Preparation of the gold colloid

We have found that gold particles larger than 8 nm penetrate poorly into the interstices of three-dimensional cytoskeletons. Larger particles are, however, useful in labelling the plasma membrane surface and the surface of Lowicryl K4M sections (36, 37, 44). Colloids of different sizes are prepared by the tannic acid methods of Slot and Geuze (44). *Table 1* shows the components for preparation of gold particles. For the preparation solution A and B are warmed to 60 °C, rapidly mixed, and heated with stirring until boiling.

Table 1. Preparation of gold particles.

Gold particle size	Solution A	Solution B
8 nm	1ml 1 per cent $HAuCl_4$ 79 ml distilled water	4 ml 1 per cent trisodium citrate 15.5 ml distilled water 0.25 ml 1 per cent tannic acid 0.25 ml 25 mM potassium carbonate
5 nm	1ml 1 per cent $HAuCl_4$ 79 ml distilled water	4 ml 1 per cent trisodium citrate 13.5 ml distilled water 1 ml 1 per cent tannic acid 1 ml 25 mM potassium carbonate
2–3 nm	1ml 1 per cent $HAuCl_4$ 79 ml distilled water	4 ml 1 per cent trisodium citrate 1 ml distilled water 5 ml 1 per cent tannic acid 5 ml 25 mM potassium carbonate

5.2 Coating gold colloid with IgGs

To coat gold with IgG, the gold solution is made to pH 9.0 using 200 mM potassium carbonate. Since unstabilized gold will rapidly destroy the electrode of the pH meter, the gold solution must be stabilized with polyethylglycol (PEG), at a final concentration of 0.05 per cent before

39

adjusting the pH. In practice, defined volumes of gold solution are titrated with different amounts of 200 mM potassium carbonate buffer and stabilized and the pH determined. The determined volume of 200 mM potassium carbonate required for a pH of 9.0 is then added to 20 ml of *unstabilized* gold solution. See *Protocol 8* for further details.

Protocol 8. Coating gold particles with IgG

1. Dialyse affinity-purified antispecies IgG at approximately 1 mg/ml concentrations into 2 mM sodium borate, pH 9.0.

2. Add 120 µg IgG to 20 ml gold solution. Stir rapidly and mix for 20 min.

3. Stabilize the gold colloid by the addition of 250 µl 8 per cent BSA (IgG free) and 20 µl 5 per cent PEG.

4. Collect gold particles by ultracentrifugation: 8 nm particles at 50 000 × g for 1 h, 5 nm and 2–3 nm particles at 125 000 × g for 1 h. The resultant pellets should be loose.

5. Remove the supernatant without disturbing the pellet.

6. Resuspend the pellet in approximately 1 ml of 150 mM sodium chloride, 20 mM Tris, 1 per cent BSA, and 0.1 per cent sodium azide, pH 8.2, and remove aggregates by centrifugation in a microcentrifuge at full speed for 10 min.

7. Store suspension at 4 °C. Gold particles prepared in this manner are stable for months. IgG-coated gold particles are used at dilutions between 1:10 and 1:100.

5.3 Coating gold colloid with protein A

The pH of the gold sol is adjusted with 0.1 M sodium hydroxide to pH 6.0 as described above. 120 µg of *Staphylococcus aureus* protein A in distilled water is added to 20 ml of freshly prepared gold sol. After 5 min, BSA and PEG are added to final concentrations of 0.1 per cent and 0.05 per cent respectively, to stabilize the gold; gold is centrifuged and resuspended in the same solution. See *Protocol 9* for the immunolabelling procedure.

Protocol 9. Immunolabelling procedure

1. Wash cover slips or formvar-coated nickel grids covered with cyto-skeletons or membrane fragments with PBS–BSA solution.

2. Prepare hydrated 10 cm Petri dishes by saturating 3–4 pieces of filter paper on the bottom of each dish with PBS and covering the moistened paper with parafilm.

3. Transfer the cover slips or grids to the Petri dishes, placing them cell side up on the sheet of parafilm.

4. Use concentrations of 1–10 μg/ml of affinity-purified IgG for maximal labelling of antigens. Dilute antigens in PBS/BSA buffer. Cover each cover slip with a 1–20 μl drop of primary IgG, and incubate for 2 h at 37 °C.

5. Wash cover slips with three changes of PBS/BSA solution; 5 min each.

6. Place 25 μl of anti-primary IgG-coated gold on each cover slip and incubate for 1 h at 37 °C. Remove unbound gold using five wash cycles of PBS–BSA, followed by three washes of PBS; 5 min each.

7. Post-fix gold-labelled cytoskeletons and membranes for 10 min at room temperature with 1 per cent glutaraldehyde in 10 mM sodium phosphate buffer. Wash with distilled water containing 2 μM phalloidin, and process for electron microscopy by either rapid freezing or negative staining.

---

6. Preparation and analysis of reconstituted polymer systems

The interaction of regulatory proteins with polymers can be studied in the electron microscope with minimal perturbation. Our work has focused on the actin fibre system and its associated proteins, and the processing of actin networks is detailed below. Similar approaches could be applied to microtubules and intermediate filaments in solution.

6.1 F-actin

The three-dimensional arrangement of actin filaments in solution in the presence and absence of cross-linking, side-binding, and capping proteins has been studied in rapidly frozen specimens. For F-actin the procedure described in *Protocol 10* allows direct visualization without fixation.

Protocol 10. Analysis of F-actin suspensions by freeze-fracturing

1. Coat small glass cover slips with 1–2 μl of F-actin solution. Keep the ionic strength of the actin solution below 10 mM to avoid salt accumulation in drying, and use buffer of 2 mM magnesium chloride, 0.5 mM ATP, 0.1 mM dithiothreitol, and 1 mM Hepes to support polymerization. Maintain the actin solution in a hydrated Petri dish (see *Protocol 9*) for 10–60 min at 37 °C.

2. Cryofix the actin solution by placing the cover slip on a square pad of 1 per cent agarose slammed on to a copper block cooled with liquid

Protocol 10. *Continued*

helium. Because the actin solution is on a cover slip, it can be easily handled and mounted on the aluminium freezing tab; however, cover slips are susceptible to detachment from the agarose block during the subsequent fracturing step. Care should therefore be taken to place the cover slip on a support about 0.5 mm larger all round. When the sample is frozen, the cover slip is thus encased on all sides within the agarose block.

3. Fracture at −120 °C. Fracture the frozen actin solution before freeze-drying it to eliminate surface tension effects at the interface between air and protein solution. It is difficult to fracture a thin frozen layer, so a simple trick is used to obtain optimally fractured samples: the knife in the freeze-fracture apparatus is always mounted with a degree of tilt relative to the sample face. This allows one side of the cover slip to be fractured near the glass surface and the other to receive a minimal cut. The ideal fracture point lies between the two extremes.

4. Freeze-dry the fractured solution for 10 min at −98 °C.

5. Metal coat, preferably with tantalum−tungsten (*Figure 3*).

---

7. Quantitation of immunogold label

The analysis of cytoskeletal structures in three dimensions and of the distribution of immunogold therein is a complex task. Three-dimensional information is obtained by taking paired micrographs at varying tilt angles in the electron microscope. Both micrographs are then used to reconstruct Z-axis information using a computer (45). A detailed discussion of some aspects of this analysis has been presented previously (27, 45). General features of gold distribution in a cytoskeleton that are to be considered without a detailed analysis include the following:

● Which type of filament is associated with the gold particles.
   The fibre systems can be identified in two ways:
   (a) *Fibre diameter and substructure.* Microtubules have diameters of 25 nm, actin and intermediate filament diameters of 10–12 nm. Although similar in diameter, the substructure of actin and intermediate fibres differs in metal-coated samples. Actin filaments have a defined substructure and helix twist of 37 nm, while intermediate filaments are smooth and untwisted. If the fibre type remains in doubt, actin and intermediate filaments can be distinguished using myosin subfragment 1. This reagent binds only to actin filaments, decorating them uniformly in twisted cable-like structures. If myosin S1 is used, it is best to apply it after the cytoskeleton has been fixed and unreacted aldehydes blocked, as it will displace many actin filament-associated proteins.

John H. Hartwig

(b) *Double immunogold labelling* using antibodies against individual filament components.

- Where along the fibre the gold is associated.
  Possibilities include ends, along sides, or at points of intersection.

- Whether immunogold is membrane-associated.
  In delipidated cells, membrane association must be inferred from the proximity of immunogold to the cytoskeletal periphery and filament ends. Substantiation requires immunogold localization in membrane-cytoskeletal fragments prepared from cleaved cells.

- Whether immunogold is aggregated or dispersed.
  Should gold label be treated as single particles or as groups of particles (clusters)? If care is taken each time before using colloidal gold probes, aggregates of gold particles in stock solutions will have been eliminated. Therefore, a finding of gold clusters in labelled specimens can be interpreted in only one of two ways: clustering of antigens, or multiple epitope labelling if the protein being probed is elongated *in situ*. In these cases, total gold, gold in clusters, and average cluster size should be determined.

7.1 Comparison between metal replicas and gold labelling in cell sections and cytoskeletal sections

The difficulty in quantifying immunogold in three dimensions requires that gold distribution also be analysed in conventionally sectioned cells (20, 46, 47, 48). We have found that embedding in the water-soluble Lowicryl K4M matrix provides both excellent morphology and high antigenicity when specimens are not exposed to osmium. Although it is not possible to resolve filamentous structures (except microtubules) at resolutions comparable to those in metal replicas, the distribution of a protein's antigenicity across cells can be easily quantified in sectioned cells, i.e. gold particle distribution in different regions of the cell can be analysed. We and others have compared the distribution of a cytoskeletal protein between the plasma membrane, other membrane-bound organelles, the nucleus, and the cytoplasm.

It is important to consider the size of the immunogold probe before defining boundaries between these various compartments. IgG molecules have dimensions of approximately 5×10 nm. Therefore, the effective size of a gold probe is the radius of the probe plus this IgG coating. Since immunolabelling with protein probes directly coupled to gold yields poor labelling, the second layer of IgG imposed on the system in the gold coating adds another 5–10 nm to the working size of the particle. Therefore, gold probes directly labelling specific structures can be functionally displaced by a maximum of about 20 nm from the structure. When quantitating integral membrane proteins, it is necessary to include all particles that lie ± 20 nm from the middle of the lipid bilayer. If membrane-associated proteins are under

43

consideration, then this distance should begin at the cytoplasm–membrane interface. When comparing membrane-bounded compartments, particles per µm of membrane length should be analysed. If the density of particle labelling is being compared between membrane and cytoplasmic compartments, then particles per µm need to be determined.

References

1. Brown, S., Levinson, S., and Spudich, J. (1976). *J. Supramol. Struct.*, **5**, 119.
2. Estes, J., Selden, L., and Gershman, L. (1981). *Biochemistry*, **20**, 708.
3. Dancker, P., Low, I., Hasselbach, W., and Wieland, T. (1975). *Biochim. Biophys. Acta*, **400**, 407.
4. Wieland, T. (1986). In *Peptides of poisonous* Amanita *mushrooms*, p. 69. Springer, New York.
5. Schiff, P., Fant, J., and Horwitz, S. (1979). *Nature* (Lond.), **277**, 665.
6. Hartwig, J. and Kwiatkowski, D. (1991). *Cur. Opin. Cell Biol.*, **3**, 87.
7. Laemmli, U. (1970). *Nature* (*Lond.*), **227**, 680.
8. Towbin, J., Staehelin, T., and Gordon, J. (1979). *Proc. Natl. Acad. Sci., USA*, **76**, 4350.
9. Small, J. (1981). *J. Cell Biol.*, **91**, 695.
10. Capco, D. and Penman, S. (1983). *J. Cell Biol.*, **96**, 896.
11. Heuser, J. and Kirschner, M. (1980). *J. Cell Biol.*, **86**, 212.
12. Schliwa, M., von Blerkom, J., and Porter, K. (1981). *Proc. Natl. Acad. Sci., USA*, **80**, 5147.
13. Glauert, A., Dingle, J., and Lucy, J. (1962). *Nature* (*Lond.*), **196**, 953.
14. Bangham, A. and Horne, R. (1962). *Nature* (*Lond.*), **196**, 952.
15. Nakata, T. and Hirokawa, N. (1987). *J. Cell Biol.*, **105**, 1771.
16. Safer, D., Golla, R., and Nachmias, V. (1990). *Proc. Natl. Acad. Sci., USA*, **87**, 2536.
17. Harding, C., Heuser, J., and Stahl, P. (1983). *J. Cell Biol.*, **97**, 329.
18. Brown, D., Gluck, S., and Hartwig, J. (1987). *J. Cell Biol.*, **105**, 1637.
19. Clarke, M., Schatten, G., Mazia, D., and Spudich, J. (1975). *Proc. Natl. Acad. Sci., USA*, **72**, 1758.
20. Hartwig, J., Chambers, K., and Stossel, T. (1989). *J. Cell Biol.*, **109**, 467.
21. Baumeister, W., Buckenberger, R., Engelhardt, H., and Woodcock, C. (1986). *Ann. N.Y. Acad. Sci.*, **483**, 57.
22. Bachmann, L. (1962). *Naturwissenschaften*, **49**, 34.
23. Bachmann, L., Abermann, R., and Zingsheim, H. (1969). *Histochemie*, **20**, 133.
24. Heuser, J. (1983). *J. Mol. Biol.*, **169**, 155.
25. Heuser, J. (1989). *J. Elect. Mic. Tec.*, **13**, 244.
26. Buckley, I. and Porter, K. (1975). *J. Microsc.* (Oxf.), **104**, 107.
27. Hartwig, J. and Shevlin, P. (1986). *J. Cell Biol.* **103**, 1007.
28. Anderson, T. (1951). *Trans. N.Y. Acad. Sci.*, **13**, 130.
29. Wolosewick, J. and Porter, K. (1979). *J. Cell Biol.*, **82**, 114.
30. Mesland, D. and Spiele, H. (1983). *J. Cell Sci.*, **64**, 351.
31. Schapp, B. and Reese, T. (1982). *J. Cell Biol.*, **94**, 667.

32. Landis, D. and Reese, T. (1983). *J. Cell Biol.*, **97**, 1169.
33. Bridgeman, P. and Reese, T. (1984). *J. Cell Biol.*, **99**, 1655.
34. Linner, J., Livesey, S., Harrison, D., and Steiner, A. (1986). *J. Histochem. Cytochem.*, **34**, 1123.
35. Jesaitis, A., Buescher, E., Harrison, D., Quinn, M., Parkos, C., Livesey, S., and Linner, J. (1990). *J. Clin. Invest.*, **85**, 821.
36. Roth, J., Brown, D., Norman, A. W., and Orci, L. (1982). *Am. J. Physiol.*, **243**, F243.
37. Roth, J. (1982). In *Techniques in Immunochemistry*. Bullock, G. R., (ed.), p. 107. Academic Press, New York.
38. Wolosewick, J. and De Mey, J. (1982). *Biol. Cell.*, **44**, 85.
39. Fey, E., Krochmalnic, G., and Penman, S. (1986). *J. Cell Biol.*, **102**, 1654.
40. Lin, A., Krockmalnic, G., and Penman, S. (1990). *Proc. Natl. Acad. Sci., USA*, **87**, 8565.
41. Nickerson, J., Krockmalnic, G., He, D., and Penman, S. (1990). *Proc. Natl. Acad. Sci., USA*, **87**, 2259.
42. De May, J., Moeremans, M., Geuens, G., Nuydens, R., and DeBrabander, M. (1981). *Cell Biol. Int. Rep.* **5**, 889.
43. Faulk, W. and Taylor, G. (1971). *Immunocytochemistry*, **8**, 1081.
44. Slot, J. and Geuze, H. (1985). *Eur. J. Cell Biol.*, **38**, 87.
45. Niederman, R., Amrein, P., and Hartwig, J. (1983). *J. Cell Biol.*, **96**, 1400.
46. Hartwig, J., Chambers, K., Hopica, K., and Kwiatkowski, D. (1989). *J. Cell Biol.*, **109**, 1571.
47. Hartwig, J. and DeSisto, M. (1991). *J. Cell Biol.*, **112**, 407.
48. Hartwig, J., Brown, D., Ausiello, D., Stossel, T., and Orci, L. (1990). *J. Histochem. Cytochem.*, **38**, 1145.

3

Actin filament assembly and organization *in vitro*

JOHN A. COOPER

1. Introduction (refs 1, 2, and 3)

Actin filaments are essential components of the contractile apparatus of muscle, and contribute to the control of shape and movement of non-muscle cells. The filaments are helical polymers of 42 kD subunits, connected by non-covalent bonds. Subunits are in dynamic equilibrium with the filaments, adding and leaving only at the two ends of the filaments, which are described as *barbed* or *pointed* based on the basis of the direction of arrowheads observed when filaments are decorated with a fragment of myosin (*Figure 1*). Because interactions are restricted to the ends, the assembly is described as linear, and the subunit concentration never rises above a value defined by the quotient of the reaction's dissociation rate constant (k_-) divided by the association rate constant (k_+) (Equation 1). In addition, the assembly is described as nucleated because small oligomers, such as dimers and trimers,

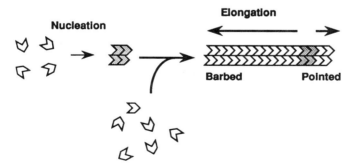

Figure 1. Actin polymerization. Subunits, drawn as chevron-shaped monomers, associate with each other to form nuclei. Further addition of subunits generates filaments, which have barbed and pointed ends. Elongation, or filament growth, is favoured at the barbed as opposed to the pointed end. All of the reactions drawn in the forward direction also occur in the reverse direction

47

are relatively unstable. Therefore, solutions at steady state include long filaments and subunits, not oligomers or short filaments. Also, no filaments are observed unless the total actin concentration exceeds a critical concentration, c_1, that also equals k_-/k_+:

$$c_1 = k_-/k_+ \qquad (1)$$

Actin polymerization is regulated *in vitro* by solution constituents and conditions (*Table 1*) and by proteins that bind to actin. Moderate ionic strength favours polymerization, as does a slightly acidic pH. Each subunit binds one molecule of nucleotide, generally ATP, and a divalent cation, generally Ca^{2+} or Mg^{2+}, which are important for stability. In addition, ATP favours polymerization more than ADP (4) and Mg^{2+} favours polymerization more than Ca^{2+} (5, 6). Among the actin-binding proteins, some bind to monomers and thereby shift the equilibrium away from filaments. Some bind to ends and prevent subunit addition and loss. Others bind to the side of filaments and regulate actin polymerization, regulate interactions of actin with other proteins such as myosin, or cross-link the filaments into networks or bundles (2, 3).

Table 1. Effect of solution conditions on actin polymerization. Actin generally requires both a divalent cation and a nucleotide to be stable

	Favours polymer	**Favours monomer**
pH	low (5–6)	high (8–9)
Ionic strength	moderate (0.1 M NaCl)	low (zero)
Nucleotide	ATP	ADP
Divalent cation	Mg^{2+}	Ca^{2+}
Temperature	high (37 °C)	low (4 °C)

This chapter includes a selection of simple assays that have been used to purify actin-binding proteins and to characterize actin polymerization and the effects of other molecules, including actin-binding proteins, on actin polymerization. Some of these assays and some other assays have been discussed previously (7, 8). Original papers, some of which are referenced, include further details on these assays, modifications of these assays, or other, more sophisticated assays that are beyond the scope of this chapter.

2. Assays with purified components *in vitro*

This section includes assays that measure the polymerization and organization or physical state of purified actin *in vitro*. The assays are organized by technique, not by functional application.

2.1 Simple rheology assays

These assays measure the viscosity of solutions of purified actin filaments, which reflects their interaction with actin-binding proteins. The apparent viscosity of these solutions is proportional to the number, concentration, and length of the actin filaments (the product of which is the weight concentration). In general, monomer-binding proteins lower the apparent viscosity by sequestering monomers away from filaments, and end-binding proteins (sometimes called capping proteins) lower the apparent viscosity by promoting the formation of shorter filaments. Side-binding proteins that cross-link the filaments can either raise the viscosity if a network is formed or lower the viscosity if the filament network collapses into bundles. Some cross-linking proteins can either raise or lower the viscosity, depending on their concentration.

These assays are generally useful only for screening large numbers of samples, such as column fractions during purification, and for preliminary characterization of actin-binding activities. The assays are not linear, so defining the centre of a peak of activity or distinguishing two peaks often requires repeating the assay with different amounts of material.

The assays require purified actin, which can be prepared from rabbit or chicken skeletal muscle (9). 'Conventional' or 'Spudich–Watt' actin, prepared by a cycle of polymerization–depolymerization (10) contains trace amounts of the barbed-end capping protein CapZ (11), which lowers the apparent viscosity. This actin preparation is often suitable for detecting the activity of proteins that increase the apparent viscosity, but may not be suitable for detecting the activity of proteins that decrease the apparent viscosity. In this case, the actin can be further purified by gel-filtration (12) to remove the CapZ. Actin prepared by either method can be stored lyophilized with sucrose at −20 °C (13, 14).

2.1.1 Falling (rolling) ball viscometry

This assay, developed by Pollard and co-workers (15), measures the speed with which a small steel ball rolls down a glass capillary that contains a solution of actin filaments. The capillary is held at an angle in a simple home-made plastic apparatus (8, 15). The actin that is mixed with the test material can be either monomeric or filamentous. If monomeric, the actin polymerizes during the incubation. Although the two approaches should yield the same result at long times, one generally wishes to incubate the samples for a short time. If one is testing a monomer-binding protein, the actin should probably be in the monomeric form. On the other hand, we find that starting with actin filaments makes the results more reproducible and allows the use of lower actin concentrations, which makes the assay more sensitive to activities that lower the apparent viscosity, such as capping proteins. A typical assay is presented in *Protocol 1*.

Protocol 1. Falling ball viscometry assay

Materials

• Buffer G: 2 mM Tris/HCl, 0.2 mM ATP, 0.2 mM CaCl$_2$, 0.5 mM NaN$_3$, 0.1 mM dithiothreitol, pH 8.0. A 50 × stock of buffer G without calcium chloride can be stored at −20 °C.

Method

1. Prepare actin (9, 11) in either monomeric form in Buffer G or polymerized at 25 μM for 10 min at room temperature by the addition of 20–100 mM salt, such as NaCl or KCl, and/or 1–2 mM MgCl$_2$. Lowering the pH to 7 or even 6 favours polymerization, as does chelation of Ca^{2+} with EGTA and addition of Mg^{2+}.

2. Dilute the actin filaments to 10 μM with polymerization buffer, mix well and incubate for 15 min at room temperature. This dilution step makes the subsequent pipetting of the actin more reproducible.

3. For each sample to be assayed, mix actin, at a final concentration of 2–10 μM, with the test material in polymerizing buffer. The total volume of each sample should be 200 μl.

4. Immediately load the sample into a capillary tube (e.g. 100 μl micro-pipette, VWR Scientific), seal one end with SealEase (Clay Adams), and incubate in a water bath at 20 °–37 °C for 10–60 min. The apparent viscosity should increase with increased actin concentration, incubation time, and temperature.

5. Drop a small steel ball (e.g. 0.64 mm diameter, material 440–C, grade 10, gauge −0.000075″, New England Miniature Ball Co.) onto the meniscus. Place the tube in the plastic holder apparatus (8, 15) at one of several possible angles, chosen by estimating the apparent viscosity.

6. Push the ball through the meniscus with a wire and measure the time required for it to roll a certain distance. Calculate inverse speed (s cm^{-1}), which can be converted to apparent viscosity using standard solutions of glycerol in water (8, 15).

2.1.2 Tube tipping assay

This assay is easier than the falling ball assay, especially for assays of column fractions, but can require more actin and does not yield a number for apparent viscosity.

The assay is the same as the previous one except that the 0.2 ml sample contains 5–25 μM actin in a 10 × 75 mm glass test tube. The sample is not loaded into a capillary, but simply incubated at room temperature for 20–60 min. The tube is then tipped on its side, and a note is made of how readily the solution flows away from the bottom.

2.1.3 Ostwald (capillary) viscometry

In this assay, a fixed volume of solution flows through a long glass capillary. The time is inversely proportional to the viscosity. The solution is sheared to a degree that actin filaments break, so the assay is not useful for characterizing networks of actin filaments prepared with cross-linking proteins. The measured viscosity is sensitive to the weight, concentration, and length of the actin filaments, which has led to its use in characterizing proteins that restrict the length of filaments and in measuring the time course of actin polymerization from monomers. Better assays now exist for both of these applications, and are described below. Details on performing this assay are described in ref. 7.

2.2 Quantitative assays for extent of polymerization

2.2.1 Introduction and miscellaneous assays

The assays of choice to measure the weight concentration of actin polymerization (without contributions from filament number and length) use one of three available fluorescent derivatives of actin—pyrene, NBD, or rhodamine. The advantage of these assays is the linear relationship between fluorescence and polymer concentration. The disadvantages are preparation of the fluorescent derivatives and requirement for a fluorimeter. Alternatives that use unlabelled actin are measurements of light scattering and absorbance at 232 nm. These assays also require expensive equipment, and have other drawbacks that make them less desirable. They have been described previously (7) and will not be discussed further here.

2.2.2 Sedimentation of actin filaments

Another alternative that uses unlabelled actin is sedimentation, which requires an ultracentrifuge, preferably one that accepts small samples, such as an Airfuge or TL100 (Beckman Instruments). In this assay, actin filaments are sedimented, and protein assays are performed on the supernatant and pellet to determine the amount of polymer. The assay is particularly useful to detect binding of proteins to actin filaments (16). Using radiolabelled protein or antibodies to the protein, the assay can be sensitive on a small scale and provide quantitative binding information (17). In this case, one needs to include a negative control, with a substance that does not bind to actin, to adjust for the volume of solution trapped in the pellet of actin filaments. To lessen this non-specific trapping, one can overlay the actin mixture on to a cushion of 10 per cent sucrose in the centrifuge tube and then pellet the filaments through the cushion. The actin concentration should be well above the critical concentration, 2–20 µM. Centrifuge an Airfuge-size tube (200 µl) for 15–30 min at > 100 000 × g. The pellet and supernatant should be analysed by SDS-PAGE to determine that the actin filaments sediment as expected. In particular, if filaments are suspected to be short, the centrifugation time should be increased and the supernatant checked for short filaments by electron microscopy.

2.2.3 Assays using fluorescent derivatives of actin

The following assays use fluorescent derivatives to measure actin filament concentration and thereby determine parameters related to actin polymerization.

i. Preparation and properties of fluorescent derivatives of actin

Three fluorescent derivatives of actin, the fluorescence of which changes on polymerization, have been described. All the procedures involve modification of Cys 374 of actin, which can be specifically labelled with thiol-reacting groups because Cys 374 is far more reactive than the other four Cys residues. One procedure also modifies an adjacent residue (Lys 373). Derivatives at this location are probably sensitive to polymerization because this site is involved in actin–actin interactions in the filament (18, 19). As with any derivative, one must consider whether the labelled actin behaves in a manner similar to the unlabelled actin. For rhodamine and NBD actin, the labelled actin polymerizes slightly faster than unlabelled actin (20), although in separate experiments such an effect was not observed for pyrene actin (6, 21). This problem can often be avoided by using actin that contains only a trace (0.1–10 per cent) of pyrene actin. In addition, certain actin-binding proteins do not bind normally to labelled actins, as mentioned below. A control experiment to test for this difference can be performed with different ratios of labelled to unlabelled actin.

We generally use conventional actin for the labelling reaction with the fluorescent dye. The reaction mixture is centrifuged at high speed to collect actin filaments, dialysed into buffer G to depolymerize the actin, recentrifuged at high speed, and gel-filtered over Sephadex G150 as usual to remove traces of CapZ (11) and free dye.

Pyrene actin, described by Kouyama and Mihashi (22), can be prepared by reacting *N*-(1-pyrenyl) iodoacetamide (Molecular Probes) with actin filaments prepared in the absence of a thiol (21, 22, 23). The shape of the fluorescence spectra changes during polymerization, which allows selection of wavelengths at which the fluorescence signal increases up to 20-fold (excitation wavelengths 365 nm, emission wavelength 388 or 407 nm) (21, 22, 23). The fluorescence is sensitive to photobleaching, so excitation must generally be minimized by either narrowing the excitation slit, adding neutral density filters, or opening the excitation shutter for brief periods of time. An additional consideration is that the fluorescence of pyrene actin is sensitive to the nucleotide bound to actin. Actin binds one molecule of nucleotide, which is required for activity.

Monomers generally have bound ATP, which is hydrolysed to ADP shortly after the monomer is incorporated into a filament (4). The fluorescence of pyrene actin with ADP is greater than that with ATP, so that when polymerization is very rapid, the time course of pyrene fluorescence does not agree with that of polymerization (24). NBD actin, described below, does not show this phenomenon.

Profilin does not bind to pyrene actin as well as it does to unlabelled actin in some experiments (25, 26) but not others (27, 28). Heavy meromyosin does not bind as well to pyrene actin as it does to unlabelled actin (22).

NBD-labelled actin is prepared by first reacting actin with N-ethylmaleimide, which reacts with Cys 374, and then with 7-chloro-4-nitro-2,1,3-benzox-adiazole (Molecular Probes), which reacts with Lys 373 (29). Fluorescence emission increases 2.2-fold on polymerization (excitation wavelength 480 nm, emission wavelength 520 nm) (29) and is not sensitive to ATP hydrolysis (24).

Rhodamine-labelled actin can be prepared by reacting actin filaments with iodoacetamidotetramethyl rhodamine (Molecular Probes) (30). On polymerization, the fluorescence spectra do not change in shape, but the fluorescence emission increases about 1.5-fold (excitation wavelength 550 nm, emission wavelength 570 nm) (30). The nucleotide sensitivity is not known.

ii. Determination of critical concentration at steady state

The critical concentration for polymerization, which is the same as the monomer concentration at steady state in a solution with filaments, can be determined from an experiment in which the total concentration of actin is varied in a range around the critical concentration. Using a fluorescent derivative of actin, the graph of fluorescence vs. actin concentration shows an inflection point at the critical concentration (*Figure 2*). Samples with actin at a range of concentrations above and below the estimated critical concentration are prepared under the conditions to be tested and incubated to steady state. The fluorescence of each sample is then measured. The time to steady state can be long, up to hours, so one needs to repeat the measurement until it is clear that steady state has been reached.

iii. Measurements of assembly and disassembly at ends

Actin filaments grow or shrink by the addition or loss of subunits at their ends when the monomer concentration is greater than or less than the critical concentration, respectively. Using fluorescent actin to monitor the weight concentration of polymer, one can design assays in which actin filaments and monomers are used to determine the number of free filament ends, the critical concentration of those filament ends, or the monomer concentration.

Depolymerization by dilution

In this assay, a sample of actin filaments is diluted to a large extent, which

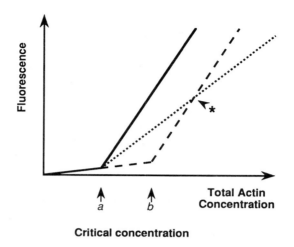

Figure 2. Determination of critical concentration at steady state. Fluorescence of an actin derivative is plotted against total actin concentration. At low actin concentrations, only monomer is present and the slope is therefore low. At high actin concentrations, above the critical concentration, the slope is high because filaments are present and increments of total actin contribute to the filament concentration only. The critical concentration is the bend in the curve, the point at which filaments first appear. The solid line represents a control experiment, for which the critical concentration is *a*. The dashed line represents a condition, such as addition of profilin, in which the critical concentration is increased to *b*. The dotted line represents another condition in which the critical concentration is not changed from *a*, but the fluorescence of the filaments is decreased. In this case, the filaments are present in the same concentration as in the solid line, but the specific fluorescence is decreased. With only a single measurement, such as at *, one would not know whether the dashed or dotted line was correct. These data are hypothetical.

favours depolymerization from filament ends. The rate of deploymerization reflects the number of ends capable of depolymerizing. Therefore, this assay can be used to determine (a) the number of filaments, if one knows that the ends are not capped, or (b) whether filament ends are capped. For example, as hypothetically illustrated by *Figure 3*, the addition of a barbed-end capping protein, such as CapZ, to actin filaments markedly slows the rate of deploymerization because the barbed ends cannot deploymerize and the dissociation rate constant of the pointed end is much less than that of the barbed end (14). However, addition of a severing protein, such as gelsolin, increases the rate of depolymerization because new filament ends are created (31). Even though gelsolin remains bound to the new barbed end created by the cut, the increase in number of pointed ends can be sufficient to increase the overall depolymerization rate (32). A typical assay is presented in *Protocol 2*.

Protocol 2. Depolymerization by dilution assay

1. Polymerize pyrene actin at a high concentration, such as 25 μM, for 30 min.

2. Dilute to 5 μM in polymerizing buffer, mix well, and incubate for 1 h to allow equilibration of monomer and polymer.

3. For each assay, remove an aliquot of 50 μl and mix the solution in a standard reproducible manner (e.g. vortex at a high setting for 30 s), which should break the filaments uniformly. Without this step, the 50 μl aliquot is usually sheared to a variable extent by the pipette.

4. Add the 50 μl sample to 2.5 ml of buffer, either buffer G (see *Protocol 1*) or a polymerizing buffer, in a fluorometer cuvette with stirring and monitor fluorescence over time. This step dilutes the actin to 0.1 μM, which is near the critical concentration of most polymerizing buffers. For a more reproducible time course, the cuvette solution may be stirred throughout.

5. An actin-binding protein may be added when the actin is diluted to 5 μM or when it is diluted to 0.1 μM.

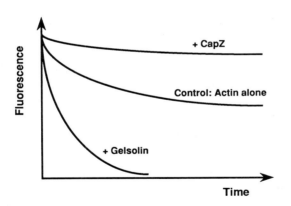

Figure 3. Effect of actin-binding proteins on depolymerization of actin filaments. Fluorescence of an actin derivative is plotted against time. Labelled actin filaments are diluted to the critical concentration at time zero. An actin-binding protein, CapZ or gelsolin, is present in the dilution buffer or in the filament mixture. CapZ slows the depolymerization rate because it binds to barbed ends and prevents the dissociation of actin subunits. Gelsolin severs actin filaments, thereby creating more ends, which accelerates the depolymerization rate. Gelsolin also caps barbed ends, which decreases the depolymerization rate; however, the increase in concentration of pointed ends overcomes the effect of capping barbed ends. These data are hypothetical.

Quantitative analysis of the time course is not straightforward. The time course of depolymerization may be described by a single exponential or it may be biphasic, which is not understood (32). Often a simple qualitative comparison of the curves, including appropriate positive and negative controls, is sufficient. Some investigators determine the initial slope of the original curve or the slope of a linear region of a plot of log fluorescence vs. time.

Polymerization by addition of filaments to high monomer concentrations
In this assay, actin filaments are used to seed the polymerization of relatively high concentrations of actin monomers. Filament elongation is described by equation 2, in which N is the number of filament ends and c_1 is the monomer concentration:

$$\text{rate} = N\,(k_+c_1 - k_-). \tag{2}$$

If c_1 is large, the second term, k_-, is much smaller than the first term. The rate is a linear function of both N and c_1, so the assay can be used to determine concentrations of filaments or monomers in experimental samples. For example, this approach was used to monitor the decrease in concentration of filament ends in a sample recovering from fragmentation by sonication (33, 34). Also, the approach was used to determine relative elongation rate constants in experiments where N was constant and c_1 was varied (35). This assay can be used to characterize actin-binding proteins and determine their affinity for actin. Addition of a capping protein to the filaments will decrease their apparent number (36), and addition of a monomer-binding protein, such as profilin, to the monomers will decrease their apparent concentration (25). A typical assay is presented in *Protocol 3*.

Protocol 3. Polymerization assay with filament seeds and high
              monomer concentrations

1. Prepare filaments of pyrene actin at 5 µM as for the depolymerization assay (*Protocol 2*).

2. Add pyrene-actin monomers at known concentrations from 0.5 to greater than 10 µM to polymerizing buffer in a fluorimeter cuvette.

3. Immediately add a 50 µl aliquot of the actin filaments, mixed before use in a reproducible manner, as for the depolymerization assay (*Protocol 2*).

4. Record fluorescence over time and measure the rate of the initial linear phase.

    ● A control without filaments is important to determine that the monomers alone do not polymerize at a significant rate.

● An actin-binding protein can be added to the filaments or monomers for a variable time before mixing or even at the time of mixing, depending on the desired information.

---

iv. Measurements of critical concentration before steady state

In these assays, one generally adds a small number of filaments to a low concentration of monomers and determines over a short time whether growth or shrinkage occurs. By making determinations at a range of monomer concentrations, one can determine the apparent critical concentration—the point at which the filaments neither grow nor shrink. A hypothetical experiment is illustrated in *Figure 4*.

The value of the apparent critical concentration can indicate whether the barbed ends are free or capped; therefore, the assay is generally used to determine whether an actin-binding protein caps barbed filament ends. In Mg^{++}, as opposed to Ca^{++}, the critical concentrations of the barbed and pointed ends are different, being about 0.1 μM and 0.5 μM, respectively. An apparent critical concentration of 0.5 μM indicates that the barbed ends are capped. The association and dissociation rate constants for subunit addition to the barbed end are usually much larger than those for the pointed end. Therefore, the apparent critical concentration of filaments with both ends free is near that of the barbed end, and if the pointed end is capped, the change in the critical concentration is probably not measurable. Therefore, the assay is not useful to characterize pointed-end capping proteins. The experiments of Selve and Wegner (37) illustrate the usefulness of this and other pre-steady state assays. A typical assay is presented in *Protocol 4*.

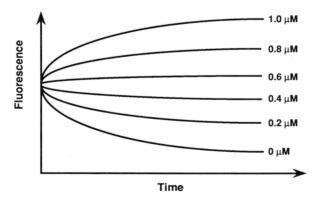

Figure 4. Determination of critical concentration from the pre-steady state time course of actin polymerization. Fluorescence of an actin derivative is plotted against time. Actin filaments are mixed with different concentrations of actin monomers, indicated on the graph. At approximately 0.6 μM, the filaments are relatively stable. At lower concentrations, the filaments depolymerize, and at higher concentrations, the filaments polymerize. These data are hypothetical.

Protocol 4. Measurement of critical concentration before steady state

1. Prepare pyrene-labelled actin filaments at 5 μM in a polymerizing buffer with 2 mM Mg^{2+} and 1 mM EGTA with and without actin-binding protein as appropriate.

2. Incubate until steady state is reached, which may be minutes to hours, depending on the protein.

3. At time zero of each assay, dilute a stock of pyrene-labelled actin monomers into a fluorimeter cuvette containing the same polymerizing buffer used above. The actin filaments and monomers should be labelled with pyrene to the same extent. The final actin monomer concentration should be approximately 0.02–2 μM.

4. Immediately add approximately 50 μl of the actin filament mixture and record fluorescence. In general, one need only determine whether the fluorescence is increasing or decreasing so a short time course is sufficient.

5. Test a range of monomer concentrations to determine at what concentration the fluorescence does not change. This is the apparent critical concentration.

v. Polymerization from monomers

When a solution of actin monomers at a concentration greater than the critical concentration is induced to polymerize by the addition of salt or Mg^{2+}, actin filaments form. The time course of polymerization can be monitored by the fluorescence of an actin derivative. The time course generally has an initial lag phase, followed by an exponential phase and a plateau. The exponential phase is addition of subunits to filament ends, as described by Equation 2 above. The lag phase is most likely accounted for by the nucleation process, in which subunits associate with each other to form small oligomers. The oligomers are relatively unstable and tend to dissociate into subunits before additional subunits are added. However, once the oligomers are large enough, they are stable and elongate as a filament.

The time course of actin polymerization from monomers has been analysed to determine the parameters of nucleation and the rate constants for the initial reaction (5, 6, 38). These studies indicate that the size of the nucleus is a trimer and that the equilibrium constants for association in the nucleation process are orders of magnitude less than those for the elongation process. Stating that the size of the nucleus is a single species, like a trimer, is an oversimplification. The nucleus is defined here as the species that has rate constants for addition of the next monomer that are equal to the elongation rate constants. Some investigators define the nucleus as one subunit larger, describing it as the smallest species that tends to polymerize more than depolymerize (38). Both definitions are meaningful, although somewhat arbitrary. The dissociation rate constant of a given species is fixed, but its association rate, at which it adds another subunit, depends on the concentration of monomer (5). Therefore, the apparent size of the nucleus varies with the monomer concentration. Also, the association reactions for species smaller than the nucleus should be less favourable than the reaction that forms the nucleus. For example, the reaction for dimer formation is less favourable than for trimer formation (5).

The experiments for this approach are straightforward and similar to those described above in this section, except that the fluorimeter cuvette contains only actin monomers at time zero. The analysis of the time course for polymerization, which allows one to infer the nucleation parameters, was done by numerical integration of rate equations based on theoretical models for the association reactions (5, 6, 38). KINSIM, the computer program used by Frieden and co-workers (39), has been developed into a user-friendly version and adapted to VAX, PC, and Macintosh computers. KINSIM is regularly used by graduate students in an introductory enzymology course, and the reader is encouraged to obtain and use it. (Dr Frieden can be contacted at the Department of Biochemistry and Molecular Biophysics, Washington University Medical School, St. Louis, MO 63110, USA, or by E-mail on BITNET at FRIEDEN@WUMS. All versions of the programs should soon be available on INTERNET, and Dr Frieden will have information about them.)

This approach has also been useful in analysing the mechanism of action of actin-binding proteins that affect polymerization (*Figure 5*). The computer-assisted analysis provides a quantitative approach to evaluating different

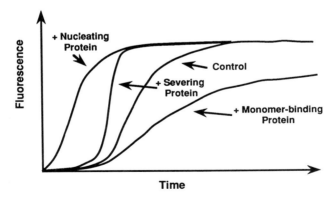

Figure 5. Effect of actin-binding proteins on the time course of actin polymerization from monomers. Fluorescence of an actin derivative is plotted against time. The control is actin alone, without additions. The other curves show the effects of addition of a monomer-binding protein, such as profilin; a nucleating protein, such as CapZ; and a severing protein, such as actophorin. These data are hypothetical.

mechanisms and providing parameters. End-binding proteins that nucleate actin polymerization, such as capping protein from *Acanthamoeba* (36) accelerate the time course and decrease the lag time. Since end-binding proteins also prevent elongation at the end to which they are bound, which is the barbed end in most cases, the time course of elongation can be slowed, as discussed in a previous section above. Paradoxically, the lag phase is lengthened by this effect; nucleation has been increased, but the new filaments are capped. This phenomenon has been demonstrated in experiments with low concentrations of macrophage capping protein (MCP) whose activity can be controlled by Ca^{2+} (40). Monomer-binding proteins, such as *Acanthamoeba* profilin (25, 41), slow the time course overall.

This approach has provided evidence that actin filaments fragment spontaneously in undisturbed solutions (6, 42). When fragmentation occurs, the time course is accelerated in the exponential phase but not the lag phase (*Figure 5*). Based on this information, further work showed that rabbit muscle tropomyosin, a protein that binds along the side of filaments, inhibits fragmentation (43) and that *Acanthamoeba* actophorin accelerates fragmentation (severs filaments) (44).

2.3 Electron microscope assays

The basic elements of electron microscopy are not described here, but Chapter 2 of this volume describes more advanced techniques.

2.3.1 *Limulus* acrosomal processes as seeds

The sperm of the horseshoe crab, *Limulus*, contain a bundle of actin filaments called the acrosomal process. Acrosomal processes are thinner at one end, and all of their actin filaments are oriented with the barbed ends at the thin end of the structure. Addition of actin monomers results in growth of new, unbundled filaments from each end, and the ends can be identified as barbed or pointed because of the thickness of the end of the acrosomal process. (*Figure 6*).

Acrosomal processes have been useful for determining the critical concentration (45) and rate constants (46) of actin polymerization at the barbed and pointed ends, as well as for determining whether actin-binding proteins prevent growth at one end or the other. The design of these assays is similar to those described in the preceding sections where actin filaments are used as seeds for actin monomer polymerization. One can either add relatively high concentrations of actin monomer and determine the rate of growth at each end, or add low concentrations, in a range surrounding the critical concentration, and determine whether growth occurs at all. The bundles are stable and do not depolymerize; therefore, observations analogous to decreasing fluorescence in the previous section are not possible.

The newly grown actin filaments splay apart from the end of the acrosomal process, so one can determine whether an actin-binding protein cross-links

John A. Cooper

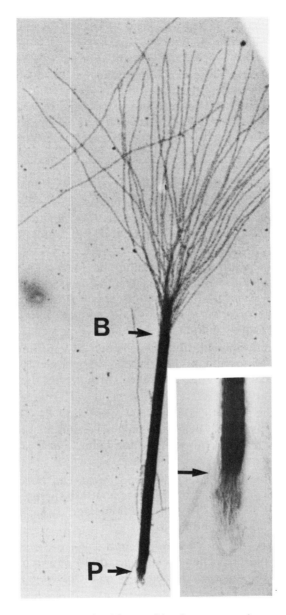

Figure 6. Actin filaments polymerized from a *Limulus* acrosomal process. This electron micrograph, kindly provided by Dr Edward M. Bonder of Rutgers University, shows an acrosomal process with actin polymerized from the barbed (B) and pointed (P) ends. The inset shows the pointed end at higher magnification.

the filaments, which have the same polarity (47). An actin-binding protein that severs actin filaments, such as villin or gelsolin, will cut the newly grown actin filaments, but not the bundle. Therefore, one can detect severing activity by adding an actin-binding protein after the new actin filaments have grown, which should cause rapid shortening of the filaments, faster than predicted by endwise depolymerization (48, 49).

The preparation of acrosomal processes from *Limulus* and a typical experiment are described in *Protocols 5* and *6*, respectively. Further details of the performance of these experiments have been described in refs 7 and 46.

Protocol 5. Preparation of acrosomal processes from *Limulus*[a]

1. Purchase male *Limulus* from the Division of Marine Resources of the Marine Biological Laboratory at Woods Hole, or from Gulf Specimens. House the animals in room-temperature salt water tanks or use upon receipt.

2. Collect sperm by turning the animal upside down (the tail and legs are not dangerous), and lifting the flap nearest the front. Sperm are released as a milky liquid from two papillae on either side of the middle either spontaneously or upon massage of the surfaces lateral to the papillae. Collect 0.5–1 ml of sperm.

3. Wash 3–4 times in cold, filtered sea water by centrifugation for 4 s in a microfuge. The sperm must be resuspended very gently—add the solution below the pellet and rock the tube. Do not pass the sperm through a pipette.

4. Resuspend the pellet by adding 1 per cent Triton X-100, 30 mM Tris/HCl, pH 8, and 3 mM $MgCl_2$ below the pellet. Loosen the pellet from the tube wall. Rotate the tube end-over-end for 2 min. This treatment releases the acrosomal processes from the sperm.

   • At this point and later, the preparation can be checked by phase contrast or differential interference contrast microscopy. Sperm heads are round to oval, flagella are long and straight, and acrosomal processes are long and helical.

5. Centrifuge for 15 s in a Microfuge and remove the supernatant, being careful to avoid the pellet. Centrifuge for 10 min in a Microfuge in the cold, discard the supernatant, and suspend the pellet in 2× polymerization buffer (as chosen for the experiment).

6. Homogenize the pellet by passing it 10–20 times through a Pasteur pipette. Re-centrifuge and wash into 2× polymerization buffer twice more. The final material can be sheared more extensively by multiple passes through a 25 gauge needle, which generates a large number of acrosomal process fragments.

- One should probably make a fresh preparation each day, although purified acrosomal processes can be stored in 50 per cent glycerol at −70 °C with variable loss of activity (50).

a This procedure was modified from those of Bonder (45, 49) and Pollard (46) by Mr Steven Heiss of our laboratory.

Protocol 6. Use of *Limulus* acrosomal processes

1. To ensure that the acrosomal processes are at an appropriate concentration and that actin assembly is linear with time, test each preparation in a pilot experiment before the main experiment.

2. In the pilot experiment, mix 10 µl of acrosomal processes in 2× polymerization buffer with 10 µl of monomeric actin in buffer G at 2 and 4 µM for 30 s.

3. Prepare a grid (below) and examine in the electron microscope. Determine (a) that the acrosomal processes are at a density convenient for observations (2–3 per grid square), (b) that nearly all acrosomal processes have actin grown from their barbed (thin) ends, and (c) that the growth in the 4 µM sample is about twice as long as that in the 2 µM sample. If many ends have not grown while some have grown extensively, the preparation can be purified by repeating the final centrifugation or a new preparation can be made. If the 4 µM sample has filaments shorter than expected, then the quantity of acromosal processes is probably too great, causing the monomer concentration to fall significantly over time. Use fewer acrosomal processes.

4. For the actual experiment, mix 10 µl of acrosomal processes with 10 µl G-actin.

5. Choose concentrations and incubation times based on the goal of the experiment. An actin-binding protein can be added before or after the actin.

6. At the end of the incubation, prepare a sample for the microscope by quickly touching a glow-discharged, carbon- and Formvar-coated grid to the sample drop and then transferring it to a larger (200 µl) drop of 1× polymerization buffer.

7. Quickly remove excess fluid by touching the grid to filter paper and then placing the grid on a drop of 1 per cent uranyl acetate for negative staining. Remove excess fluid again, dry in air and then under vaccum before examining the grid in a transmission electron microscope.

   - The 1× polymerization buffer can be supplemented with 5 mM spermine, pH 7, which bundles the newly grown actin filaments and makes it easy to measure their lengths on the microscope (46). Without

Protocol 6. *Continued*

spermine, one generally photographs the acrosomal processes and measures lengths on the photographic negative or print.

• If the acrosomal processes are sheared to make them shorter, which makes it easier to find both ends of one fragment, then distinguishing the two ends often requires measuring the thickness of the two ends. The end of the acrosomal process can be distinguished from the newly grown actin filaments because the acrosomal process has a crosshatched pattern to its appearance, owing to the protein that bundles the filaments.

An alternative seed for these experiments, in place of the *Limulus* acrosomal process, is a short filament of actin decorated with heavy meromyosin. The heavy meromyosin will dissociate from the actin filament in the presence of ATP, so if ATP is to be used in the polymerization buffer (which is nearly always the case), the seeds must be cross-linked with glutaraldehyde and gel-filtered. The preparation protocol is described in ref. 7.

2.3.2 Electron microscopy for filament length determination and bundle formation

Measurement of filament length remains a difficult problem. Solution techniques, such as quasi-elastic light scattering (51) and fluorescence photobleaching recovery (52) yield an average value for a population of filaments. The filaments must be fully mobile and therefore relatively short and/or dilute. One study found similar lengths by quasi-elastic light scattering and electron microscopy (51). Electron microscopy affords accurate length measurements of individual filaments; however, sample preparation may break long filaments and the length of the filament may affect the efficiency with which the filament adheres to the grid. Sample preparation is performed by diluting the filaments to an appropriate concentration, near 0.5 µM, applying them to a grid and negative-staining, as described above for the *Limulus* acrosomal processes. Bundles of actin filaments can be readily observed; therefore, this technique is useful for characterizing actin-binding proteins that cross-link filaments. Other types of sample preparation, which are more difficult but provide more information about the filaments, are discussed in Chapter 2 by Hartwig.

2.4 Optical microscope assays

The basic techniques of optical microscopy are not described here, but Chapter 1 by Wang describes methods used for microscopy of living cells.

2.4.1 Visualization of filament bundles

Sample preparation for electron microscopy may alter the supramolecular organization of the filaments. Optical microscopy does not have this disadvantage because samples can be prepared and examined in a single chamber. Phase contrast, differential interference contrast, and polarization microscopy can detect bundles of a sufficient size with unlabelled actin. With fluorescent-labelled actin, fluorescence microscopy can detect inhomogeneities in filament concentration and distribution, in addition to bundles.

2.4.2 Fluorescence photobleaching recovery (FPR)

FPR (also called fluorescence recovery after photobleaching, or FRAP) is useful to characterize the motions of actin filaments, which can yield information about their size and organization, and to characterize the binding of proteins to actin filaments. While the assembly and use of an instrument for FPR is beyond the capability of many investigators, this section discusses some applications of FPR for the investigator who wishes to perform experiments in a specialized laboratory with a suitable instrument. Quasi-elastic light scattering, which is not discussed here, can also characterize the motions of filaments, and yield information about their physical state (51, 53). Some studies have used these techniques to examine the motion of inert particles trapped in networks of actin filaments to characterize the networks (54, 55, 56, 57). Quantitative rheology, which is beyond the scope of this discussion, also yields information about networks of filaments. Viscometers for these studies generally apply a minimal shear that oscillates over time to yield information about elasticity and viscosity (53, 56, 58–62).

In FPR the fluorescence of a labelled protein is determined. The light source is focused by an optical microscope on to a spot or bar, so the fluorescent signal, recorded by a photomultiplier tube attached to the microscope, reflects the concentration of labelled protein in that small region of the solution. A laser is used to provide high light intensity to the spot for a brief time, during which many of the fluorescent molecules are bleached. After the bleach, the fluorescence signal is correspondingly decreased. However, if molecules from outside the bleached spot diffuse into the spot, the signal will increase. The time course of the recovery of the signal can be analysed to determine the translational diffusion coefficient of the fluorescent molecule. A typical experiment is shown in *Figure 7*.

This approach has been useful for characterizing the motions of actin filaments and microtubules (52). If the solution is dilute, so that the diffusion of one molecule is not inhibited by the presence of the other molecules, then the diffusion coefficient can be interpreted in terms of the size, specifically the length, of the filaments (52). Only one value is obtained, which reflects the whole population of filaments, so information about the distribution of lengths among the population is not available.

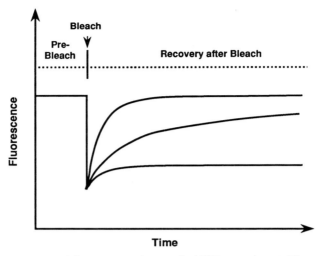

Figure 7. Time course of fluorescence in a typical FPR experiment. Three hypothetical curves are shown. Before the bleach, the three curves overlap. After the bleach, the curves describe full recovery of a fast molecule (upper), full recovery of a slower molecule (middle), and partial recovery, i.e., the presence of an immobile fraction (lower). Long actin filaments often show an immobile fraction.

FPR can also be used to characterize the binding of labelled proteins to unlabelled actin filaments. At high concentrations, the filaments move very slowly, because they either collide with each other or have weak interactions with each other. For whichever reason, on the time scale of most FPR measurements, the filaments are immobile and show no fluorescence recovery. In this case, one can analyse the binding of a fluorescently-labelled actin-binding protein. Unbound protein will diffuse at its normal rate, determined from a control experiment, and bound protein will not diffuse at all. Therefore, the fraction of protein that is bound can be determined from the apparent mobility. This approach has been used to analyse the binding of α-actinin to actin (63).

3. Assays with cell preparations and extracts

Chapter 1 by Wang discusses observations of living cells and immuno-fluorescence staining of fixed cells. This section discusses observations of actin polymerization using cell extracts.

3.1 Measurements of F- and G-actin

3.1.1 Fluorescent phalloidin binding (64, 65, 66)

Phalloidin binds tightly to the actin subunits in filaments but not monomers; therefore, the amount of bound phalloidin reflects the amount of actin filaments (67). A limitation of the assay is that actin-binding proteins may inhibit the binding of phalloidin, so the measurement may not reflect all cellular actin. Fluorescent derivatives of phalloidin are commercially available (Molecular Probes). Cells are usually fixed and always permeabilized to allow fluorescent phalloidin access to the cytoplasm. Individual cells in suspension can be analyzed directly in a fluorescence-activated cell sorter (FACS). For groups of cells, either on a surface or in suspension, the fluorescent phalloidin can be extracted and measured in a fluorimeter. A fluorescence microscope equipped with a photomultiplier tube or a video camera with a digital processor can determine the fluorescence of single cells or regions of cells (68, 69). A typical assay is presented in *Protocol 7*.

Protocol 7. Fluorescent phalloidin binding assay.

This version of the assay was used to measure the fluorescence of a population of cells on a substrate with fluorimeter (70). Cells in suspension can be treated in a similar manner with washes by centrifugation and analysis with a FACS or a fluorimeter (64, 65, 66).

1. Fix the cells with 0.1–3 per cent formaldehyde in PBS for 10–30 min.

2. Permeabilize with 0.1 per cent Triton X-100 or another non-ionic detergent in PBS for 5 min.

3. Rinse and incubate with a fluorescent derivative of phalloidin, such as rhodamine-phalloidin, in PBS for 15 min. Fluorescent phalloidin is usually supplied as a solution in methanol. Remove the desired quantity and dry it with a stream of nitrogen or argon, then redissolve in PBS. The concentration and volume of the fluorescent phalloidin to be used should be sufficient to saturate the cellular actin and therefore should be determined in a pilot experiment.

4. Wash the cells quickly several times with PBS.

5. Extract the cells with either absolute methanol or 0.1 N NaOH and measure the fluorescence in a fluorimeter at appropriate wavelengths, given in the Molecular Probes catalogue.

6. In control experiments, include a 10-fold excess of unlabelled phalloidin (Sigma) to determine the level of non-specific binding, which should then be subtracted from the total to yield specific binding (69).

3.1.2 Inhibition of DNase I by G-actin

This method, described in detail previously (7, 64, 71, 72), relies on the observation that actin monomers bind tightly to DNase I and inhibit its hydrolytic activity. The assay has been used to measure G-actin and total actin in crude cell extracts. The total actin is determined by depolymerizing the actin filaments. F-actin is then calculated as total actin minus G-actin. A potential problem with the assay is that DNase I may also bind to actin filaments, although the location is controversial (68, 69), and DNase I can depolymerize actin filaments (73), perhaps by mass action.

Interpretations of data from this assay should consider that complexes of actin with certain actin-binding proteins, such as profilin, also bind to DNase I (74). Therefore, the value for G-actin obtained does not reflect that of free actin monomers. Also, the behaviour of membrane-associated actin (75) is unpredictable in this assay. The G-actin/total actin fraction as measured in this assay is near the value determined by high-speed centrifugation (e.g. 72)). A typical assay is presented in *Protocol 8*.

Protocol 8. Inhibition of DNAse I by actin. Typical assay, based on refs 64, 71, 72

1. Lyse the cells in 1 per cent Triton X-100, 2 mM $MgCl_1$, 1 mM EGTA, 50 mM Tris/HCl, pH 7.5. Determine whether the actin is stable in the lysis buffer: for example, Fox *et al.*, using platelets, found that a similar buffer without Mg^{2+} worked well but that addition of ATP made the actin filaments unstable (72).

2. Add 20 μl cell lysate to 20 μl DNase I (EC 3.1.21.1, Type II, from bovine pancreas) at 0.1 mg/ml in 50 mM Tris/HCl, pH 7.5, 0.1 mM $CaCl_2$, 10 μM PMSF (phenylmethylsulfonylfluoride).

3. Add the mixture to 3 ml of DNA (calf thymus DNA, Type I, Sigma) at 80 μg/ml in 0.1 M Tris/HCl, 4 mM $MgSO_4$, 1.8 mM $CaCl_2$ at 25 °C. To dissolve the DNA, cut the fibrous material into small pieces with scissors, and stir or rotate slowly for several hours.

4. Continually monitor the absorbance at 260 nm, which increases as the DNA is hydrolysed.

5. Determine the rate of absorbance increase in the linear portion or the inflection point of the time course.

6. Use varied amounts of purified monomeric actin in place of cells to construct a standard curve.

7. To determine total actin, add 20 μl of cell lysate to 20 μl of 1.5 M guanidine HCl, 1 M Na acetate, 1 mM $CaCl_2$, 1 mM ATP, 20 mM Tris/HCl, pH 7.5, and incubate for 5–30 min at 4 °C. Then add the DNase I and proceed as above. The guanidine solution may inhibit the DNase I activity (64), so repeat the actin standard curve with guanidine.

3.2 Nucleation of actin polymerization by cell extracts

In this assay, cell extracts or preparations, such as whole cell lysates or membrane fractions, are added to fluorescent actin monomers. Conditions are chosen so that the actin monomers nucleate slowly, and the cell extract accelerates the nucleation rate. One generally determines the initial linear rate of fluorescence increase, which should be proportional to the number of nuclei. The extract may induce the formation of filaments that are free at their barbed or pointed ends, which can be distinguished by the use of inhibitors. Cytochalasin D or a barbed-end capping protein, such as CapZ, will inhibit barbed-end growth, and the spectrin/actin/band 4.1 complex of erythrocytes will inhibit pointed-end growth (76–78). Cytochalasin D is commercially available, but the protein complexes are not. One needs to perform controls without lysate and with cytochalasin or the actin-binding protein to ensure that the latter agents do not nucleate on their own.

Acknowledgements

I am grateful to Dr Edward Bonder for assistance with the description of the *Limulus* acrosomal process assay and for providing Figure 6, and to James Amatruda and Steven Heiss for reading the manuscript. Preparation of this chapter was funded by grants from NIH and the Lucille P. Markey Charitable Trust. The author is a Lucille P. Markey Scholar.

References

1. Alberts, B., Bray, D., Lewis, J., Raff, M., Roberts, K., and Watson, J. D. (1989). *Molecular biology of the cell*, p. 613. Garland Publishing, New York.
2. Pollard, T. D. (1990). *Curr. Opin. Cell Biol.*, **2**, 33.
3. Pollard, T. D. and Copper, J. A. (1986). *Ann. Rev. Biochem.*, **55**, 987.
4. Korn, E. D., Carlier, M. F., and Pantaloni, D. (1987). *Science*, **238**, 638.
5. Frieden, C. (1983). *Proc. Natl. Acad. Sci. USA*, **80**, 6513.
6. Cooper J. A., Buhle, E. L., Jr., Walker, S. B., Tsong, T. Y., and Pollard, T. D. (1983). *Biochemistry*, **22**, 2193.
7. Cooper J. A. and Pollard, T. D. (1982). *Meth. Enzymol.*, **85**, 182.
8. Pollard, T. D. and Cooper, J.A. (1982). *Meth. Enzymol.*, **85**, 211.
9. Pardee, J. D. and Spudich, J. A. (1982). *Meth. Enzymol.*, **85**, 164.
10. Spudich, J. A. and Watt, S. (1971). *J. Biol. Chem.*, **246**, 4866.
11. Casella, J. F. and Maack, D. J. (1987). *Biochem. Biophys. Res. Commun.*, **145**, 625.
12. MacLean-Fletcher, S. M. and Pollard, T. D. (1980). *Biochem. Biophys. Res. Commun.*, **96**, 18.
13. Frieden, C. (1982). *J. Biol. Chem.*, **257**, 2882.
14. Caldwell, J. E., Heiss, S. G., Mermall, V., and Cooper, J. A. (1989). *Biochemistry*, **28**, 8506.

15. MacLean-Fletcher, S. D. and Pollard, T. D. (1980). *J. Cell Biol.*, **85**, 414.
16. Nunnally, M. H., Powell, L. D., and Craig, S. W. (1981). *J. Biol. Chem.*, **256**, 2083.
17. Pope, B. (1986). *Eur. J. Biochem.*, **161**, 85.
18. Kabsch, W., Mannherz, H. G., Suck, D., Pai, E. G., and Holmes, K. C. (1990). *Nature*, **347**, 37.
19. Holmes, K. C., Popp, D., Gebhard, W., and Kabsch, W. (1990). *Nature*, **347**, 44.
20. Tait, J. G. and Frieden, C. (1982). *Biochemistry*, **21**, 6046.
21. Tellam, R. and Frieden, C. (1982). *Biochemistry*, **21**, 3207.
22. Kouyama, T. and Mihashi, K. (1981). *Eur. J. Biochem.*, **114**, 33.
23. Cooper, J. A., Walker, S. B., and Pollard, T. D. (1983). *J. Mus. Res. Cell Motil.*, **4**, 253.
24. Carlier, M.-F., Pantaloni, D., and Korn, E. D. (1984). *J. Biol. Chem.*, **259**, 9983.
25. Lal, A. A. and Korn, E. D. (1985). *J. Biol. Chem.*, **260**, 10132.
26. Malm, B. (1984). *FEBS Lett.*, **173**, 399.
27. DiNubile, M. J. and Southwick, F. S. (1985). *J. Biol. Chem.*, **260**, 7402.
28. Lee, S., Li, M., and Pollard, T. D. (1988). *Anal. Biochem.*, **168**, 148.
29. Detmers, P., Weber, A., Elzinga, M., and Stephens, R. E. (1981). *J. Biol. Chem.*, **256**, 99.
30. Tait, J. G. and Frieden, C. (1982). *Arch. Biochem. Biophys.*, **216**, 133.
31. Kwiatkowski, D. J., Janmey, P. A., and Yin, H. L. (1989). *J. Cell Biol.*, **108**, 1717.
32. Walsh, T. P., Weber, A., Higgins, J., Bonder, E. M., and Mooseker, M. S. (1984). *Biochemistry*, **23**, 2613.
33. Pantaloni, D., Carlier, M.-F., Coue, M., Lal, A. A., Brenner, S. L., and Korn, E. D. (1984). *J. Biol. Chem.*, **259**, 6274.
34. Carlier, M.-F., Pantaloni, D., and Korn, E. D. (1984). *J. Biol. Chem.*, **259**, 9987.
35. Pollard, T. D. (1983). *Anal. Biochem.*, **134**, 406.
36. Cooper, J. A. and Pollard, T. D. (1985). *Biochemistry*, **24**, 793.
37. Selve, N. and Wegner, A. (1986). *Eur. J. Biochem.*, **160**, 379.
38. Tobacman, L. S. and Korn, E. D. (1983). *J. Biol. Chem.*, **258**, 3207.
39. Barshop, B. A., Wrenn, R. F., and Frieden, C. (1983). *Anal. Biochem.*, **72**, 248.
40. Young, C. L., Southwick, F. S., and Weber, A. (1990). *Biochemistry*, **29**, 2232.
41. Pollard, T. D. and Cooper, J. A. (1984). *Biochemistry*, **23**, 6631.
42. Wegner, A. and Savko, P. (1982). *Biochemistry*, **21**, 1909.
43. Wegner, A. (1982). *J. Mol. Biol.*, **161**, 217.
44. Cooper, J. A., Blum, J. D., Williams, R. C., and Pollard, T. D. (1986). *J. Biol. Chem.*, **261**, 477.
45. Bonder, E. M., Fishkind, D. J., and Mooseker, M. S. (1983). *Cell*, **34**, 491.
46. Pollard, T. D. (1986). *J. Cell Biol.*, **103**, 2747.
47. Fishkind, D. J., Mooseker, M. S., and Bonder, E. M. (1985). *Cell Motil.*, **5**, 311.
48. Harris, H. E. and Weeds, A. G. (1984). *FEBS Lett.*, **177**, 184.
49. Bonder, E. M. and Mooseker, M. S. (1983). *J. Cell Biol.*, **96**, 1097.
50. Bonder, E. M. and Mooseker, M. S. (1986). *J. Cell Biol.*, **102**, 282.
51. Janmey, P. A., Peetermans, J., Zaner, K. S., Stossel, T. P., and Tanaka, T. (1986). *J. Biol. Chem.*, **261**, 8357.
52. Doi, Y. and Frieden, C. (1984). *J. Biol. Chem.*, **259**, 11868.

53. Janmey, P. A., Hvidt, S., Oster, G. F., Lamb, J., Stossel, T. P., and Hartwig, J. H. (1990). *Nature*, **347**, 95.
54. Zaner, K. S. and Valberg, P. A. (1989). *J. Cell Biol.*, **109**, 2233.
55. Newman, J., Schick, K. L., and Zaner, K. S. (1989). *Biopolymers*, **28**, 1969.
56. Zaner, K. S., King, R. G., Newman, J., Schick, K. L., Furukawa, R., and Ware, B. R. (1988). *J. Biol. Chem.*, **263**, 7186.
57. Newman, J., Mroczka, N., and Schick, K. L. (1989). *Biopolymers*, **28**, 655.
58. Sato, M., Schwarz, W. H., and Pollard, T. D. (1987). *Nature*, **325**, 828.
59. Sato, M., Leimbach, G., Schwarz, W. H., and Pollard, T. D. (1985). *J. Biol. Chem.*, **260**, 8585.
60. Janmey, P. A., Hvidt, S., Peetermans, J., Lamb, J., Ferry, J. D., and Stossel, T. P. (1988). *Biochemistry*, **27**, 8218.
61. Zaner, K. S. and Hartwig, J. H. (1988). *J. Biol. Chem.*, **263**, 4532.
62. Zaner, K. S. (1986). *J. Biol. Chem.*, **261**, 7615.
63. Simon, J. R., Furukawa, R. H., Ware, B. R., and Taylor, D. L. (1988). *Cell Motil. Cytoskeleton*, **11**, 64.
64. Howard, T. H. and Meyer, W. (1984). *J. Cell Biol.*, **98**, 1265.
65. Howard, T. H. and Oresajo, C. (1985). *J. Cell Biol.*, **101**, 1078.
66. Meyer, W. H. and Howard, T. H. (1987). *Blood*, **70**, 363.
67. Cooper, J. A. (1987). *J. Cell Biol.*, **105**, 1473.
68. Southwick, F. S., Dabiri, G. A., Paschetto, M., and Zigmond, S. H. (1989). *J. Cell Biol.*, **109**, 1561.
69. Cassimeris, L., McNeill, H., and Zigmond, S. H. (1990). *J. Cell Biol.*, **110**, 1067.
70. Dadabay, C. J., Patton, E., Cooper, J. A., and Pike, L. J. (1991). *J. Cell Biol.*, **112**, 1151.
71. Blikstad, I., Markey, F., Carlsson, L., Persson, T., and Lindberg, U. (1978). *Cell*, **15**, 935.
72. Fox, J. E. B., Dockter, M. E., and Phillips, D. R. (1981). *Anal. Biochem.*, **117**, 170.
73. Hitchcock, S. E., Carlsson, L., and Lindberg, U. (1976). *Cell*, **7**, 531.
74. Carlsson, L., Nystrom, L.-E., Sundkvist, I., Markey, F., and Lindberg, U. (1977). *J. Mol. Biol.*, **115**, 465.
75. Carraway, K. L., Cevra, R. F., Jung, G., and Carraway, C. A. C. (1982). *J. Cell Biol.*, **94**, 624.
76. Lin, D. C. and Lin, S. (1979). *Proc. Natl. Acad. Sci. USA*, **76**, 2345.
77. Lees, A., Haddad, J. G., and Lin, S. (1984). *Biochemistry*, **23**, 3038.
78. Cribbs, D. H., Glenney, J. R., Kaulfus, P., Weber, K., and Lin, S. (1982). *J. Biol. Chem.*, **257**, 395.

4

Mapping structural and functional domains in actin-binding proteins

PAUL MATSUDAIRA

1. Introduction

At the molecular level, two endpoints of a protein structure/function study are the protein sequence and three-dimensional structure. However, crystallography and sequence determination have not yet reached the utility and simplicity of other, less direct, methods like microscopy and spectroscopy. As a result, less powerful but faster methods must be used in the interim. Protein modification methods are arguably the most important set of experimental tools for elucidating the structure and function of a protein. Generally, these methods fall into two categories. The first group truncates the protein into a smaller piece either by enzymatic or chemical cleavage or through expression of mutant proteins deleted in various regions of the DNA coding sequence. The second group of methods uses enzymes, chemicals, or site-directed DNA mutagenesis to modify a residue in the protein into a chemical derivative or another amino acid. Through a carefully planned combination of both approaches, one can quickly identify not only the structural domains that play an important role in the function of the protein, but also the residues that are critical in binding ligand molecules or target proteins. Although DNA-based methods are very powerful, many considerations suggest that protein-chemical studies should be a first step in investigating the structure and function of a protein. The most important consideration is that in the native structure, the sites of peptide cleavage and amino acid modification are restricted to residues that lie at the surface of the protein or are at least accessible to the solvent. Simple experiments can be devised to test the importance of each site for the normal function of the protein. Once these functionally important regions are located in the sequence of a protein, then DNA mutagenesis of the region can be applied in an efficient and focused manner. This chapter outlines protein-chemical approaches to elucidating the domain organization of actin-binding proteins by describing an integrated set of methods (4–7) that my laboratory has adopted in the analysis of cytoskeletal proteins of the intestinal brush border.

The ultimate goal of many studies is to identify the binding sites of a ligand molecule or protein and to study how binding interactions are regulated. Realizing that such sequences comprise a small fraction of the total sequence, we are faced with a difficult task without knowledge of the protein sequence or the three-dimensional structure. However, the search is simplified by a divide-and-conquer approach in which the protein is first cleaved into smaller fragments, either by chemical methods or by proteolysis, and then the activity of each fragment is tested. Binding sites become much easier to identify in an active fragment of a protein than in the intact protein. Thus, initial studies of a protein might be profitably directed toward mapping the domain organization of the protein. Once domains are identified and mapped, their actin-binding properties can be studied. This minimal information provides a limited view of the structure and function of the protein.

2. Chimeric structure and function of actin-binding proteins

The structures of many actin-binding proteins are chimeric, consisting of several domains which carry out separate, characteristic functions. A good example of a multidomain, multifunctional protein is the heavy chain of muscle myosin. Myosin is an actin motor protein which is responsible for movements associated with actin filaments (2). In the electron microscope, myosin is seen as a two-headed molecule with a rod-shaped tail. The N-terminal half of myosin comprises the head, while the C-terminal half forms an α-helical coiled coil. The head domain, when separated from the tail by limited proteolysis, displays the characteristic actin-activated ATPase activity of the parent molecule and can support ATP-dependent movement of actin filaments in *in vitro* motility assays. The coiled tail domain self-assembles into a thick filament that functions to anchor the force-transducing myosin heads to the thick filament backbone of muscle.

Other actin-binding proteins having different activities also consist of separate functional domains. For example, rod-like actin cross-linking proteins aggregate a solution of actin filaments into tangles of bundles or networks (3). Globular actin-binding domains lie at the ends of the rod, and the rod forms by self-association through repeating α-helical or β-sheet secondary structure elements. In addition to the oligomerization domains, actin cross-linking proteins often have additional but separate membrane and calcium-binding domains.

Actin-severing proteins (1), including villin, gelsolin, fragmin, and severin, are representative of a different type of multidomain protein. These proteins fragment actin filaments into shorter segments. One end of each segment becomes 'capped' by the actin-severing protein, which prevents actin monomers or actin filaments from binding to that end. Severing and capping activities require calcium. These proteins are built up from a common domain that is repeated three times in fragmin and severin and six times in villin and gelsolin. Severing probably involves interactions between the domains, because individual domains of villin and gelsolin that are generated by limited proteolysis have lost severing activity and display either actin capping or F-actin-binding activity. Severing proteins are good models for illustrating practical approaches for mapping domains and locating important functional regions such as actin binding sites.

3. Mapping the domain structure of a protein

3.1 Structural and functional domains

A domain can be defined in two ways. From a structural perspective, a domain is a region of compactly folded polypeptide. Small proteins ($<$ 20 kDa) are often composed of a single domain, while large proteins usually consist of multiple domains. For large proteins, domains can be recognized in structures determined by X-ray crystallography or in images captured by electron microscopy of metal shadowed or negatively-stained samples. In the three-dimensional structure of a protein a domain is seen as a unit of independently folded polypeptide that is well differentiated from other parts of the protein. In contrast to this structural definition, a second view of a domain is that it is the smallest piece of a protein that retains function. Functional regions are identified by proteolysis or by mutagenesis experiments in conjunction with an assay for function. Functional domains usually consist of one or more structural domains.

3.2 Identifying structural domains

Traditionally domains are identified by proteolysis (8); this procedure remains the method of choice because it offers the advantages of quickness, simplicity, and sensitivity. As illustrated in *Figure 1*, sites for proteolytic cleavage in a native protein can be grouped into three classes: accessible, moderately accessible, and inaccessible. Accessible sites lie on the surface of a folded protein where they are most exposed. Moderately accessible sites are secondary sites of cleavage that are exposed once a protein has been digested and the fragments have dissociated. Inaccessible sites lie in the protein interior where proteases are excluded from cleaving the polypeptide. In denatured proteins all sites along the polypeptide chain are accessible. However, certain flanking residues may inhibit cleavage at a specific site.

The amount of protease in the digest mixture is the most important variable that determines whether cleavage occurs between domains or within domains of a native protein. In the presence of small quantities of protease, cleavage is restricted to the most accessible sites, usually short stretches of polypeptide

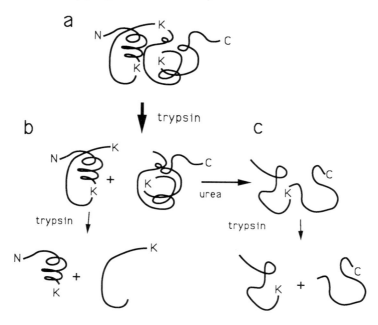

Figure 1. Accessibility of lysine (K) residues in a folded protein to protease. (a) Lysine residues on the outer surface of the protein will be the most accessible, while the internal lysines will be protected from proteases. (b) As digestion proceeds, the domains dissociate and expose additional sites. (c) Sites within a folded domain are accessible only when the domain is unfolded by denaturants

that connect domains or surface loops of the polypeptide. In the presence of large amounts of protease, conditions that are normally used to digest the protein into its smallest fragments, exposed regions of the protein are cleaved rapidly, which enables quicker access to more internal sites in the protein.

A procedure for cleaving proteins with proteases is outlined in *Protocol 1*. It is crucial that proteolysis is performed under conditions in which the protein has native structure. The amount of protease used to identify domains is determined by trial and error. Different ratios of protease:protein (weight for weight) are assayed by SDS-PAGE to identify the number and sizes of fragments released at different times of digestion. The total digestion time is kept to 1 h for convenience. The concentration of the protein is held fixed, while the protease concentration is varied. As a general rule, the concentration of protease should be kept as low as possible so that within a 1 h period of digestion, the most accessible sites are cleaved first and secondary cleavage products appear later. Characteristically, only a few fragments are generated at early times of digestion. The number of fragments increases and the size of the fragments decreases w h longer times of digestion.

Protocol 1. Limited proteolysis of native proteins

Choice of protease

Perform proteolysis with proteases that cleave specifically after acidic, basic, or hydrophobic residues. All of the proteases listed here cleave at the C-terminal peptide bond.

- Trypsin (EC 3.4.21.4) cleaves after the basic residues Arg and Lys.
- Endoproteinase Arg-C (submandibular proteinase A: EC 3.4.21.40) or endoproteinase Lys-C: EC 3.4.99.30) is used if more restricted cleavage after Arg or Lys is desired.
- Endoproteinase Glu-C (*S. aureus* protease V-8: EC 3.4.21.9) cleaves after the acidic residues Glu and Asp. By avoiding phosphate buffers, cleavage can be made to occur specifically after Glu.
- Chymotrypsin A (EC 3.4.21.1) cleaves specifically after aromatic residues Tyr, Phe, and Trp.

Preparation of proteins and proteases

1. Solubilize the protein at a 1–5 mg/ml concentration in non-denaturing aqueous solutions. Pipette 50 μl solution into a 150 μl polypropylene microcentrifuge tube.
2. Solubilize the proteases at 1–5 mg/ml concentration. The final concentration is chosen so that addition of the protease will minimally (< 10 per cent) dilute the digest mixture. Keep the concentrated protease solution on ice and use < 3 h after preparation.
3. Prepare beforehand tubes in which the reactions will be quenched, e.g. add aliquots (0.5 μl) of PMSF (0.1 M in isopropanol) to 150 μl microcentrifuge tubes.

Proteolysis

1. Cleave the undenatured protein by addition of protease. The digestion (50 μl) is performed with a 1:200–1:5000 protease:protein ratio (w/w) at room temperature.
2. Withdraw aliquots (10 μl) from the digest mixture at 1, 4, 20, and 60 min of digestion and quench by pipetting the aliquot into prepared tubes containing PMSF.
3. Dissolve the samples by addition of 2 μl of concentrated (5×) SDS sample buffer made according to the Laemmli formula (17). Denature the protein and protease by boiling in water for 4 min (in a microwave oven).

Protocol 1. *Continued*

SDS-PAGE

1. Load the digest mixture on to 20-well mini-gels (0.5 mm thick). 2–5 µg of total digest in a lane is usually sufficient to detect the major proteolytic fragments. Thin mini-gels offer the advantages of being fast and requiring small amounts of sample (21). Two gel formulas are used. Fragments > 10 kDa are separated on gels containing a gradient of polyacrylamide and using the Tris–glycine buffer system of Laemmli (22). The range of polyacrylamide concentrations in the gradient is chosen based on the size of the intact protein and its smallest fragments. Typically 5–20 per cent gels are used to resolve 200 kDa proteins from their proteolytic fragments, 5–15 per cent gels for 100 kDa proteins, and 10–20 per cent gels for 50 kDa proteins. Proteins and peptides < 20 kDa are separated on 14 per cent acrylamide gels electrophoresed in the Tris-tricine buffer system of Schagger and von Jaglow (23). This gel system resolves peptides as small as 5 kDa.

2. Visualize polypeptides by staining with Coomassie Blue. A minor band on a thin mini-gel represents 10–20 ng. To detect very small quantities of peptides, the gels should be stained using silver enhancement methods.

Analysis

The protein should be cleaved to small fragments in a slowly progressing fashion within a 1 h period. Typically, some intact protein is present after 4 min of digestion, during which intermediate-size fragments begin to accumulate. By 20 min the undergraded protein is no longer detected and intermediate-sized fragments appear. The fragments should be of discrete sizes. A smear of poorly resolved bands usually suggests that the protein is unfolded. Later, small (10–30 kDa) protease-resistant fragments accumulate.

A typical time course of digestion of villin with several proteases is shown in *Figure 2*. Early on, intact villin is still detected in the digest reaction, but with time there is a progressive decrease in sizes of the proteolytic fragments. Analysis of the kinetics of the appearance of the fragments shows that villin is cleaved nearly in half by trypsin into two fragments of M_r 44 and 51 kDa. Similar sized fragments are generated by chymotrypsin or V-8 protease (endoproteinase Glu-C), but only after villin is initially cleaved into an 87 kDa and a 10 kDa fragment. Without sequence or antibody cross-reactivity information, it is impossible to order these fragments in the polypeptide.

Comparisons of the cleavage patterns generated by several proteases is important in identifying domains. *Figure 2* shows that trypsin, endoproteinase Glu-C, and chymotrypsin all generate fragments of similar size during the

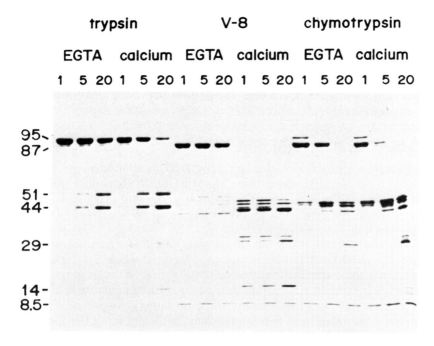

Figure 2. Time course (1, 5, and 20 min) of villin digestion with trypsin, V-8 protease (Endoproteinase Glu-C), and chymotrypsin in EGTA and calcium. The relative molecular masses of the major proteolytic fragments are shown on the left. Trypsin initially cleaves villin into 51 kDa and 44 kDa fragments. Later, 29 kDa and 14 kDa fragments are detected. V-8 protease and chymotrypsin both cleave villin into an 87 kDa core fragment and an 8.5 kDa headpiece fragment. The core is cleaved into fragments of similar size to those seen after trypsin digestion

digest, suggesting that there are common protease-sensitive and -resistant regions. At longer times of digestion fragments of 10–14 kDa begin to accumulate, suggesting that these are the end products of limited proteolysis.

3.3. Calcium-induced conformation changes

Conformational changes in the three-dimensional structure of a protein can be detected by differences in the peptide map generated in the presence and absence of a ligand (8). The actin-severing activity of villin is activated by calcium (9). In the absence of actin, calcium is able to induce changes in the difference spectra and hydrodynamic properties of villin (10, 11), which in turn indicate a calcium-dependent change in conformation. As shown in *Figure 2*, villin is cleaved more quickly by all three proteases in calcium than in EGTA. Furthermore, although the pattern of fragments generated in calcium and EGTA appears largely identical, there are some minor differences in the sizes of the fragments. As documented in the next section, the increased rate at which villin is cleaved results from changes in its conformation rather than from increased activity of the protease, because there is a calcium-dependent shift in the sites of proteolysis. Although it is clear from the pattern of fragmentation that the conformation of villin changes in the presence of calcium, the pattern does not pinpoint where the changes occur.

3.4 Mapping structural domains

Once protease-resistant domains are generated by limited proteolysis, the next task is to identify where these domains lie in the intact protein. In general, mapping the locations of domains is a two-step process. In the first step, the fragments generated by proteolysis must be separated, and in the second step the isolated polypeptide must be identified. The problem is compounded by complexity in the digest mixture. From simple digest mixtures composed of a few fragments (see *Figure 2a*, lane 3), it is relatively easy to purify these fragments and to determine their locations in the protein. However, it is difficult to separate and purify all fragments from a complex digest mixture and, as a result, maps of the proteolytic fragments are usually incomplete. We have used two approaches for determining the location of cleavage sites. The first method uses antibodies specific for either the N- or C-terminus of the protein to identify by immunoblotting the fragments that retain either end of the protein (5). The second method identifies the N-terminal sequence of the fragments and is more direct (6, 7).

3.4.1 Mapping domain boundaries with antisera that recognize the N- or C-terminus of the protein: cleavage mapping

Sites of cleavage between domains can be mapped based on the approach diagrammed in *Figure 3*. An incomplete digest of a native protein will have more than one fragment containing the N- or C-terminus of the uncleaved protein, and these will differ in size. In an ideal case the digest mixture would contain a mixed population of polypeptides, each generated by random cleavage at a single site. The fragments are separated by SDS-PAGE, and the N-terminal fragments are detected by antibodies directed against the N-terminus of the protein. The size of the fragment detected by the antibody maps the distance to the cleavage site from the N-terminus. A second blot developed with antibodies directed against the C-terminus of the protein provides a complementary map of the cleavage sites. In practice, the protein is slowly cleaved under conditions that generate a mixture of peptides that arose through single and multiple cleavage. Conditions for incomplete cleavage of proteins using proteases are described in *Protocol 1*.

Antibodies specific for the N-terminal or C-terminal sequences are obtained by injecting rabbits with short synthetic peptides coupled to BSA or

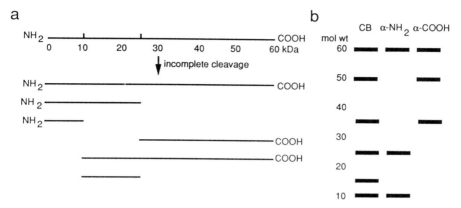

Figure 3. (a) A hypothetical 60 kDa protein contains cleavage sites located 10 and 25 kDa from the N-terminus. Incomplete cleavage of the protein generates six polypeptides including the uncleaved protein. Three fragments retain the N-terminus of the original protein, while three other fragments retain the original C-terminus. (b) The fragments are separated according to size by SDS-PAGE. An N-terminus-specific antibody recognizes three fragments and maps cleavage sites 10 and 25 kDa from the N-terminus. A C-terminus-specific antibody recognizes three fragments and maps cleavage sites to 35 and 50 kDa from the C-terminus

KLH (keyhold limpet hemocyanin). The cross-linking method is described in *Protocol 2*. The peptides are 10–15 residues in length. A C-terminal Gly–Cys sequence is included in the synthesis of the N-terminal peptide, and a Cys–Gly sequence is included at the N-terminus in the synthesis of the C-terminal peptide. The peptide is coupled to the carrier protein through the Cys residue, while the Gly residue provides a one-amino-acid spacer. If the peptide contains an internal Cys residue, then the peptide should be coupled using a residue not present in the peptide. As a second choice Tyr is substituted for Cys at the terminus of the peptide. Peptide antisera are screened for cross-reactivity with the peptide as well as the native and urea-denatured protein by ELISA or dot blot assays. Some antisera show very weak cross-reactivity with the native but not the denatured protein. This behaviour suggests that the terminus is buried in the native structure.

Protocol 2. Peptide cross-linking to carrier protein

Preparation of carrier protein–cross-linker conjugate

1. Dissolve 10 mg of carrier protein (BSA, KLH, or thyroglobulin) in 0.75 ml of phosphate-buffered saline.

2. Dissolve 1.8 mg of crosslinker *m*-maleimidobenzoyl-*N*-hydroxysuccinimide ester (MBS) (Pierce) or succinimidyl 4-(*N*-maleimidomethyl)cyclo-hexane-1-carboxylate) in 0.2 ml of dimethyl formamide (SMCC) (Pierce).

Protocol 2. *Continued*

3. Add dissolved cross-linker dropwise to a stirring solution of carrier protein. React for 30 min at room temperature.

4. Add 4.0 ml of 0.1 M $NaPO_4$, pH 6.0.

5. Separate the carrier protein–cross-linker complex from unreacted cross-linker and protein by chromatography through Sephadex G-25 equilibrated with 0.1 M $NaPO_4$, pH 6.0. Monitor the effluent at 258 nm and pool the early eluting complex.

Preparation of the peptide–carrier protein conjugate

1. Dissolve 5 mg of peptide in 1 ml of 0.1 M $NaPO_4$, pH 6.0.

2. Add dissolved peptide to carrier protein–cross-linker complex and react overnight with constant mixing at room temperature.

3. Dialyse the conjugate overnight in a 25 kDa cutoff dialysis bag against phosphate-buffered saline at 4 °C. Replace the dialysis solution once.

4. Aliquot the conjugate and store frozen at -20 °C.

If the sequence of the protein is unknown, then raising peptide antisera directed against N- or C-terminal sequences becomes problematic. In some cases the N-terminus can be sequenced directly. However, if the N-terminus is blocked, then more elaborate strategies must be used to isolate the N- or C-terminal peptides (12). Alternatively, monoclonal antibodies raised against the native protein may be fortuitously specific for the N- or C-terminus because these regions are often exposed on the surface of the protein. Such antibodies can be recognized by the correspondence between the number of fragments they recognize and the number of cleavage site residues, as determined from amino acid analysis.

Protocol 3 describes the immunoblotting procedure. *Figure 4* illustrates a cleavage blot of the villin digest shown in *Figure 2* developed with an N-terminus-specific antibody. The blot shows a series of bands representing fragments which differ by 10–15 kDa. From the sizes of the fragments, we can determine that all three proteases cleave villin at sites 14 kDa, 32 kDa, and 44 kDa from the N-terminus, and two of the three proteases cleave villin at sites 64 kDa, 75 kDa, and 87 kDa from the N-terminus. The common protease-sensitive sites delineate the boundaries of seven protease-resistant domains in villin. A map of the domains and cleavage sites is shown in *Figure 4b*. Different combinations of domains comprise proteolytic fragments that have been previously purified and characterized (13). Villin core and 44T are calcium-regulated actin-severing fragments of villin. Headpiece and 51T are F-actin binding fragments of villin that do not display actin-severing activity.

Paul Matsudaira

Protocol 3. Mapping cleavage sites with N- and C-terminus-specific antibodies (5)

1. Separate the digest by SDS-PAGE as described in *Protocol 1*.

2. During electrophoresis, equilibrate a sheet of nitrocellulose membrane (Schleicher and Schuell) with blotting buffer (10 mM 3-[cyclohexyl-amino]-1-propanesulfonic acid (CAPS), 10 per cent methanol, pH 11.0).
   - Wear gloves while handling the membrane and gel.

3. After electrophoresis, place the gel on the nitrocellulose membrane and sandwich between sheets of pre-wet 3MM paper. Place the sandwich into a tank electroblotting apparatus.

4. Transfer peptides from the gel to the membrane for 1 h at 0.5 A constant current.

5. Remove the membrane and block unbound sites with phosphate-buffered saline, pH 7.4, containing 0.1 per cent Tween 20 for 0.5 h at room temperature. BSA is not included in the blocking or antibody solutions, because the absorbed protein will contribute high background staining when the filter is stained in step 12.

6. Dilute antisera in phosphate-buffered saline, pH 7.4, containing 0.1 per cent Tween 20.

7. Incubate the membrane with the diluted antisera for 1 h at room temperature.

8. Wash unbound antibodies from the membrane by two washes of PBS followed by two washes with 0.1 per cent NP-40 in PBS and by two final washes in PBS. Each wash is of 10 min duration.

9. Incubate the membrane with a secondary antibody coupled to alkaline phosphatase for 1 h at room temperature.

10. Repeat step 8.

11. Localize the antibody by development with NBT (Nitro Blue tetra-zolium) and BCIP (*p*-nitrophenyl phosphate, disodium). The reaction is quenched by water when the bands have developed suitable colour.

12. Stain the membrane with 0.1 per cent Amido Black dissolved in 50 per cent methanol and 10 per cent acetic acid for 2 min to visualize all of the protein bands. Destain the membrane with 50 per cent methanol and 10 per cent acetic acid.

13. Calculate the relative molecular masses of the antibody-detected bands from standard samples electrophoresed in the adjacent lanes.

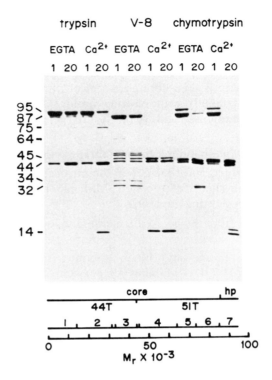

Figure 4. An immunoblot of the digest shown in *Figure 2* using an N-terminus-specific antibody. The relative molecular masses of the bands detected by the antibody allow location of the cleavage site. A map of the cleavage sites is shown below with the locations of the core, headpiece, 44T, and 51T domains indicated.

3.4.2 Mapping Cys and Met residues

The cleavage mapping method can be used to identify the locations of residues (5) if the cleavage reaction is specific for a residue. For example, the map of the domains shown in *Figure 4* is a map of the most accessible Lys, Arg, Glu, and large hydrophobic residues; however, the map is incomplete because most of these residues are buried in the protein. Cleavage is more complete in an unfolded protein where proteases will not be sterically inhibited by the secondary and tertiary structure of the protein. However, high concentrations of urea or SDS, denaturants used in completely unfolding the protein, can decrease the activity and alter the specificity of proteases (14). In addition, the residues that flank the cleavage site can sterically interfere with cleavage at some Lys and Arg residues and prevent a complete map of these residues.

These problems with reactivity are eliminated when chemical cleavage reagents are used. In contrast to proteases which cleave at abundant amino acid residues (Lys, Arg, Glu, Asp, Phe, and Tyr), chemicals such as NTCB, cyanogen bromide, and BNPS-skatole cleave only at the rarest amino acids, Cys, Met, and Trp, respectively, and thus generate fewer but larger fragments than proteases. Furthermore, cleavage (described in *Protocols 4 and 5*) is usually performed under denaturing conditions that completely inhibit proteolytic attack. Because the protein is unfolded, most, if not all, of the sites are attacked by the chemical. A trick is to devise incomplete cleavage protocols that will generate nested sets of N- and C-terminally truncated peptides. In principle, incomplete chemical cleavage methods coupled with the antibody detection scheme described in *Protocol 3* permits one to count and map these residues in the polypeptide chain. The accuracy of the map is limited by the resolution and determination of relative molecular masses of peptides by SDS-PAGE. In our experience (deArruda *et al.*, manuscript in preparation), most of the Met and Cys residues mapped within 20 residues of their predicted location. Sites (Met–Ser and Met–Thr) that are normally recalcitrant to cyanogen bromide cleavage were not detected. Maps of Cys residues are invaluable for chemical modification studies using cysteine-reactive reagents. Unfortunately, however, NTCB-treated proteins give fragments which are blocked at their amino termini and thus refractile to sequencing techniques.

Protocol 4. Mapping NTCB cleavage sites (Cys residues)

Preparation of protein and NTCB cleavage reagent

1. Reduce cysteine residues in protein with dithiothreitol (DTT) by treating the protein (0.2–2 mg) in 6 M guanidine-HCl, 10 mM DTT, 0.1 mM EDTA, 0.2 M Tris acetate (pH 8.0) at 37 °C. After 6–12 h, DTT is removed by desalting the protein into 6 M guanidine-HCl, 0.1 mM DTT, 0.1 mM EDTA, 0.2 M Tris acetate (pH 8.0).

2. Dissolve 2-nitro-5-thiocyanobenzoic acid (NTCB) (Sigma) in 50 mM sodium phosphate, pH 7.6.

Cleavage reaction

1. Add NTCB to the protein (enough for the analysis of products at six time points) at a five-fold excess over total thiols in the protein and incubate in the dark at 37 °C.

2. Remove aliquots at various times (2, 6, 12, 24, 36, and 48 h) and quench the reaction by addition of an equal volume of glacial acetic acid.

3. Dialyse the quenched cleavage mixtures against 10 per cent acetic acid (v/v) in 3500 M_r cutoff dialysis membrane (Spectrapor).

4. Dry the cleaved products by lyophilization or in a SpeedVac.

Protocol 4. *Continued*

5. Dissolve the dried fragments in SDS sample buffer at a 10 mg/ml concentration (calculated from the starting amount of protein and assuming 100 per cent recovery). Trace amounts of acid will turn the Bromphenol Blue indicator dye to a yellow colour. The pH of the mixture can be adjusted by addition of 1 μl of an 0.1 M Tris base whose pH is not adjusted. 2–5 μl of digest mixture on a mini-gel is sufficient to detect all the major cleavage fragments.

Controls

1. To identify sites cleaved by incubation in the buffer solutions, subject the reduced protein to a mock cleavage in which the NTCB is absent.

2. To identify non-cysteine cleavage sites, treat a sample of reduced and carboxymethylated protein with NTCB. Cys residues are the rarest in a protein and represent on the average 1 per cent of the total.

Protocol 5. Mapping cyanogen bromide cleavage sites (Met residues)

Preparation of protein and cyanogen bromide cleavage reagent

1. Desalt protein (0.1–1 mg) against 10 per cent acetic acid and lyophilize.

2. Prepare a 1 M solution of cyanogen bromide (Pierce) in 10 per cent acetic acid in a fume hood. Place several small crystals of cyanogen bromide in a preweighed 1.5 ml microcentrifuge tube. Cap the tube containing the crystals and reweigh. Dissolve the crystals in 10 per cent acetic acid.

   • **Caution!** Cyanogen bromide is extremely toxic and should be handled with gloves and in a fume hood. Reference 24 describes methods of inactivating and disposing of cyanogen bromide.

Cleavage reaction

1. Solubilize the dried protein in a small volume of 10 per cent acetic acid (50 μl, < 20 mg/ml). Mix cyanogen bromide with the protein (enough for analysis at five time points) at a 1000-fold molar excess over total methionine in the protein. Perform the reaction at room temperature and in the dark. As a control, leave an aliquot of protein in 10 per cent acetic acid and do not treat with cyanogen bromide. Mild acid can cause spurious cleavage of the peptide backbone, especially at Asp–Pro bonds.

2. Quench the reaction at various times (1, 6, 12, 24, and 48 h) by evaporating the mixture in a SpeedVac.

3. Dissolve the dried fragments in SDS sample buffer at a 10 mg/ml concentration (calculated from the starting amount of protein and

assuming 100 per cent recovery). If required, adjust the pH of the solubilized sample with 1 M Tris base. 2–5 µl of digest mixture on a mini-gel is sufficient to detect all of the major cleavage fragments.

4. Separate the SDS-solubilized samples by SDS-PAGE and identify the N- and C-terminal peptides as described in *Protocol 3*.

Notes

- Proteins are completely cleaved by cyanogen bromide in 70 per cent formic acid. When cleavage is performed in 10 per cent acetic acid, cleavage at Met residues is competed by oxidation at the same residues. Cyanogen bromide cleavage is inhibited at oxidized Met (18), and thus the protein is incompletely cleaved.

- The Met content is most accurately quantified by amino acid analysis. The Met content of an average eucaryotic protein is only 2.4 per cent of the total amino acids. For the purpose of calculating the amount of cyanogen bromide required for the cleavage reaction, this value can be used as a crude estimate of Met in a protein whose M_r is known although its amino acid composition is not.

3.4.3 N-terminal sequence from peptides generated by proteolysis

If the sequence of the protein is known, then the fragment is directly mapped by determining its N-terminal sequence. The sequences of the N-terminus of the most abundant fragments (1–10 pmol) are easily obtained by sequencing the digest products from blots of PVDF membranes. The blotting procedure is described in *Protocol 6*. *Figure 5* shows a blot on to PVDF membrane of separate digests of 44T and 51T, the N-terminal and C-terminal halves of villin. These fragments were purified from a trypsin digest and proteolysed with several proteases to generate smaller protease-resistant domains. The N-terminal sequences of small quantities (1–10 pmol) of peptides were easily obtained with automated protein sequencing instruments. The sequence identifies the N-terminal site of cleavage, while the C-terminal site is estimated from the size of the peptide. The map of the cleavage sites is consistent with the cleavage map calculated from immunoblots.

Protocol 6. Mapping cleavage sites by N-terminal sequence analysis (6)

1. Load the quenched digestion mixture (10–100 pmol) on to an SDS gel.

2. Follow steps 1–3 in *Protocol 3*. Transfer peptides to new generation PVDF membranes (Immobilon P^{sq}; Millipore, or ProBlott; Applied Biosystems Inc.).

Protocol 6. *Continued*

3. Transfer peptides quantitatively from the gel to the membrane for 0.2–1 h at 0.5 A constant current. The duration of the transfer is determined by the relative molecular masses of the fragments and the acrylamide concentration in the gel. Generally, 10 kDa peptides are transferred from a 10–20 per cent gel in 10 min, while 100 kDa peptides are transferred from a 5–20 per cent gel in 60 min.

4. Wash the membrane in the CAPS transfer buffer for 10 min and then in water for 10 min.

5. Visualize the blotted peptides by staining for 5 min with 0.1 per cent Coomassie Blue R-250 in 50 per cent methanol, 10 per cent acetic acid. (The bands can be excised from the blot and sequenced directly in an automated protein sequencing instrument after destaining).

6. Destain the membrane with 50 per cent methanol, 10 per cent acetic acid and wash with several changes of water.

7. Air dry the membrane and store at −20 °C.

- Peptides suitable for sequence analysis must have a minimum of 10 pmol of material in the band.

Sequence from blotted peptides also pinpointed regions in 44T that underwent a calcium-dependent change in conformation. The map shows that 44T is cleaved by trypsin into a 14 KdA and either a 27 kDa fragment in calcium or a 28 kDa fragment in EGTA. Although the latter two fragments are similar in size, their N-terminal sequences place them in different regions of the protein. The 28 kDa fragment consists of domains 1 and 2, while the 27 kDa fragment consists of domains 2 and 3. Evidently calcium induces a conformation change in 44T that prevents cleavage between domains 2 and 3. No such differences in the peptide maps of 51T were seen.

4. Assays to identify actin-binding domains

Once the domain structure of a protein is established, then strategies to characterize the functions of the domains can be planned. In preliminary studies it is useful to employ methods that identify binding domains from unfractionated digests of protein. In some cases the binding domains can be purified directly from crude mixtures by an affinity step. This approach works well with most actin-binding proteins. A key step is to take advantage of the various states of assembly in which actin filaments can exist. Differences between the states can be exploited to purify and characterize actin-binding domains.

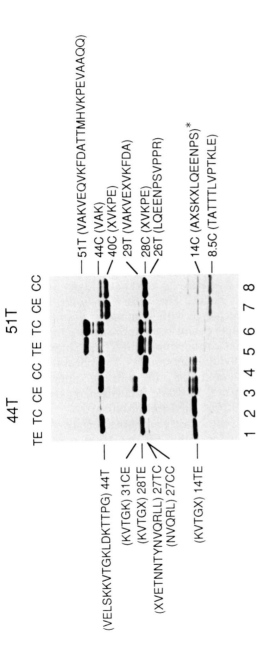

Figure 5. Digests of 44T and 51T performed with trypsin in calcium (TC) and EGTA (TE) or with chymotrypsin in calcium (CC) or in EGTA (CE) were separated by SDS-PAGE and electroblotted on to a PVDF membrane. The membrane was stained with Coomassie Blue and individual bands were sequenced in an automated gas-phase sequencing instrument. The relative molecular masses of the fragments are indicated on the sides of the blot, and the N-terminal sequences of the respective fragments are shown in one-letter code. The asterisk denotes a sequence obtained from a different blot. Based on the size of the fragment and the N-terminal sequence, the fragments were mapped to the villin sequence as shown on the diagram.

4.1 Physical states of actin

Actin can exist in three interconvertible states: a monomeric subunit, a polymeric filament, or a supramolecular bundle or network. The most significant difference among these states is size. Actin monomers measure 5.5 × 5.5 × 3.5 nm (15), while filaments of actin monomers are nearly 10 nm wide (16) and 2–10 μm long. Bundles of actin filaments, consisting of several hundred filaments cross-linked in a semi-crystalline array, measure 0.5 μm wide and 30 μm long (17). Networks, like bundles, also consist of cross-linked actin filaments but, unlike bundles, filaments in a network are arranged in a three-dimensional tangle. The dynamic interconversion between monomers and filaments and between filaments and bundles or networks forms the basis for many cell motility phenomena.

4.2 Quantitation of actin

The amount of actin that exists in each state is directly measured by simple assays that rely on the large differences in size between actin monomers and polymers. Centrifugation (see *Figure 6*) is a crude but very simple method to quantify actin. At high g-forces ($100\ 000 \times g$ for 20 min) actin filaments, but not actin monomers, sediment into a pellet. The actin remaining in the supernatant represents subunits that exchange with subunits at the filament ends. The actin concentration of the supernatant is an approximate measure of the threshold or 'critical concentration' for polymerization (see Chapter 3). Filaments will depolymerize into individual monomers at concentrations below the critical concentration. The amount of protein in the supernatant and pellet is quantified by densitometry of Coomassie Blue-stained gels or by colorimetric assays such as the Lowry, BCA, or Bradford assay.

Other methods such as fluorimetry or viscometry are more sensitive assays of actin structure and are also useful for following the kinetics of actin assembly/disassembly. These assays are described in Chapter 3 of this volume.

4.3 Actin-binding assays

Actin-binding assays measure the ability of a protein or a domain of a protein to co-sediment with actin filaments, to alter the stability or organization of actin filaments, or to alter the assembly/disassembly properties of actin. Binding is usually dependent on the ionic strength and pH of the solution, but care should be taken to devise solutions that do not affect the stability of actin filaments. For example, actin filaments depolymerize in buffers of low ionic strength, and precipitate at pH < 6.0. ATP and divalent cations (usually magnesium) are included because actin denatures in their absence. Binding buffers typically contain 25–150 mM NaCl or KCl, 1 mM $MgCl_2$, and ATP, and the pH is held at 7.0–8.0.

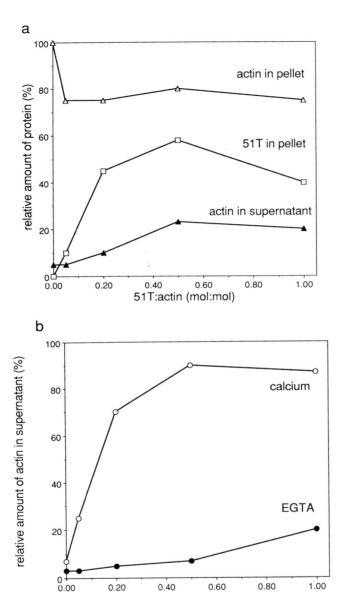

Figure 6. (a) The relative amount of actin in the supernatant (closed triangles), actin in the pellet (open triangles), and 51T in the pellet (open boxes) quantified by densitometry of SDS-PAGE gels. In the absence of 51T, all of the actin is in the pellet. 51T slightly reduces the amount of actin in the pellet and causes a reciprocal increase of actin in the pellet. Actin (10 μM) was mixed with 0–10 μM 51T in the presence of EGTA). (b) 44T in the presence of calcium (open circles) causes most of the actin to remain in the supernatant, while in EGTA (closed circles) a slight increase in actin in the supernatant is observed.

4.3.1 Actin pelleting assays

Proteins with moderate to high binding affinities ($< \mu$M Kd) for actin can be detected by simple sedimentation assays (*Figure 7* and *Protocol 7*). Actin filaments pellet at high (100 000 $\times$ g) but not low (10 000 $\times$ g) centrifugal forces. Proteins or domains that bind actin filaments will cosediment in a high g-force centrifuge, and the amount of protein in pellets and supernatants is easily quantified from SDS gels. *Figure 7* shows that the C-terminal half of villin, 51T, binds actin filaments. From quantitation of the bands in the gel, we determined that binding exhibits a 1:2 stoichiometry. The gel shows that 51T also modestly increases the amount of actin remaining in the supernatant. Binding constants and stoichiometry of binding can be determined by Scatchard analysis of the bound and unbound amounts of the protein. Care should be taken to control for unbound protein that is trapped in the water space of the pellet.

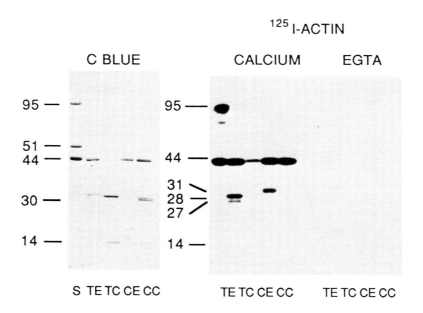

Figure 7. A gel overlay of the 44T digest products shown in a gel overlay of the 44T digest products, shown in *Figure 5*, stained with Coomassie Blue (left panel) and an autoradiogram of the gel after incubation with ^{125}I-actin (right panel). The actin binds villin, villin core, 44T, and the proteolytic fragments 28TE and 31CE. All five polypeptides retain domains one and two of villin. The 14 kDa fragments containing domain one alone do not bind actin by the overlay assay. Control experiments show that binding occurs in the presence of calcium but not in EGTA.

Protocol 7. Actin pelleting assay

Preparation

- Prepare actin from a rabbit skeletal muscle acetone powder following the protocols modified (26) from Spudich and Watt (25). Actin can be stored as F-actin (10–20 mg/ml in 30 mM NaCl, 1 mM $MgCl_2$, 10 mM TrisHCl, 0.3 mM NaN_3, pH 8.0) at 4 °C for several weeks.
- Standard pelleting assay buffer consists of 50–100 mM NaCl, 1 mM $MgCl_2$, 10 mM Pipes, pH 7.0, 0.1 mM EGTA, 0.1 mM ATP, 0.1 mM DTT.

Procedure

1. Perform the assay in 150 μl ultracentrifugation tubes (polycarbonate), if a Beckman Airfuge is available. Mix buffer and F-actin (10–20 μM final concentration) by vortexing. Because the F-actin stock solution is very viscous, measure the protein using a Hamilton syringe for accurate delivery. Add the actin-binding protein or fragment last to each tube (0, 0.2, 0.5, 2.0, 5.0, 20.0, and 50 μM final concentrations), vortex the mixture, and incubate either at room temperature or at 4 °C for 2–12 h. The final volume is 50 μl.
   - Note: If conventional ultracentrifuge is used, the total volume must be increased.
2. After incubation examine the solution for turbidity. Pipette 2 μl aliquots on to carbon films and negatively stain in 2 per cent uranyl acetate for examination by electron microscopy. Solubilize an 8 μl aliquot with an equal volume of 2× SDS sample buffer.
3. Pellet the mixture for 20 min at 100 000 × g at 4 °C.
4. Remove the supernatant; note the clarity, size, and consistency of the pellet, and resuspend the pellet with 40 μl of assay buffer; solubilize the resuspended pellet with an equal volume of 2× SDS sample buffer.
5. Analyse equal volumes of the solubilized pellets, supernatants, and uncentrifuged mixture by SDS-PAGE. The relative amounts of protein and actin are quantified by densitometry of the stained gels.

Notes

1. Pellets of F-actin are clear, circular, and very viscous. Pellets of actin bundles are turbid and smeared along the side of the tube. Furthermore, the volume of pelleted actin bundles is approximately three times larger than volumes of the equivalent amount of F-actin alone. Bundles of filaments are rigid and do not pack tightly in pellets. The pellets are not viscous, and fracture when resuspended in buffer. Using a marker of the aqueous phase to measure the volume trapped in pellets, I estimate that 2

Protocol 7. *Continued*

or 6 per cent of the initial protein is trapped in pellets of filaments or bundles, respectively. A correction for trapping may be necessary when calculating the amount of a weak affinity protein bound to actin filaments. For this reason, pelleting assays are unsuitable for proteins with $K_d > 10$ μM.

2. Actin-severing proteins will fragment actin filaments into short pieces. In the standard pelleting assay, the severed filaments will not pellet and all of the protein is recovered in the supernatant.

Highly enriched preparations of actin-binding proteins or domains can be obtained from a crude mixture by co-sedimentation with actin filaments or by velocity sedimentation (see Chapter 6). If several species pellet with the filaments, then they must be purified and individually tested for actin binding activity to identify possible cooperative interactions or to eliminate non-specific interactions by one or more of the peptides. More extensive considerations of sedimentation of actin-associated proteins are given in Chapters 3 and 6.

Pelleting assays can also reveal the actin-binding properties of the protein or domain. Macroscopic gels or bundles of actin filaments, formed by actin cross-linking proteins, sediment at low *g*-forces. The presence of actin bundles is indicated by a large, opaque pellet that fractures easily when resuspended. In addition, bundled actin filaments scatter light and do not pack tightly in a pellet. In contrast, a pellet of unbundled filaments is small and clear, and has a jelly-like consistency. Unlike actin cross-linking proteins, actin-severing or capping proteins or domains decrease the amount of actin in pellets. This is demonstrated for 44T in *Figure 6b*. The amount of actin in pellets depends on the 44T:actin molar ratio, because the stoichiometry determines the average length of the severed filament. For example, in 1:5 mixtures of 44T:actin, the actin filaments will be roughly 5 subunits long. Small complexes will remain largely in the supernatant during centrifugation. The effects of capping proteins are more difficult to detect by centrifugation assays. Usually there is a modest 2–3-fold increase in the supernatant actin in 1:1 mixtures of protein to actin. This increase reflects the increase in critical concentration when the fast assembly end of an actin filament is capped. Fluorescence-based assays described in Chapter 3 are more suitable for detecting capping activity.

4.3.2 Actin overlay assays

High-affinity interactions with actin monomers can be detected by an overlay-type assay (19) (*Protocol 8*) in which fragments of an actin binding protein are separated by SDS-PAGE and then the gel is soaked in a dilute solution of

^{125}I-actin (20). The actin is most likely monomeric, because the actin concentration is below the critical concentration for filament formation. However, because the solution contains moderate concentrations of salt, the actin monomers may have a filament-like conformation. In the case of proteins that bind actin in a calcium-dependent manner, it is advisable to perform the assay with either calcium or EGTA in the buffers. In our studies on villin and villin fragments (*Figure 7*), we found that the overlay assay mapped an actin-binding activity within the N-terminal two domains. From other experiments, we know that villin peptides identified by the gel overlay assay also display actin-severing activity.

Protocol 8. Actin gel overlay assay (20)

Preparation of buffers and protein

1. Dialyse G-actin or resuspend F-actin in actin-depolymerizing buffers that do not contain free amines. Dilute G-actin (50 µg G-actin 1 mg/ml in 2 mM Pipes, 0.2 mM $CaCl_2$, 0.1 mM ATP, pH 7.0) to 100–200 µl (> 0.5 mg/ml) with 0.1 M borate buffer, pH 8.5.

2. Prepare 2 l of gelatin-PBS buffer (100 mM NaCl, 3 mM NaN_3, 10 mM $NaPO_4$, 1 mM $MgCl_2$, 0.2 mM ATP, 1 mM DTT, 0.1–0.25 per cent gelation, pH 7.4).

Preparation of ^{125}I-actin

1. Pipette ^{125}I-Bolton and Hunter reagent (50 µCi in benzene) into a small glass vial and evaporate the benzene with a stream of dry nitrogen gas, following the safety precautions outlined by the manufacturer. Cool the dried reagent on ice.

2. Add the G-actin solution to the vial. Incubate the protein on ice for 15 min with occasional agitation.

3. Quench the reaction with 1 M glycine in 0.1 M borate, pH 8.5 (0.1 ml/2 mCi) and incubate for 5 min on ice.

4. Separate the labelled protein from the free label by desalting through Sephadex G-25 (1 ml bed volume) equilibrated with 0.1 per cent gelatin in phosphate-buffered saline (gelatin–PBS). Pool fractions containing the protein and dialyse against gelatin–PBS buffer overnight. Typical yields are 5–15 per cent incorporation, assuming 100 per cent recovery of protein. The estimated specific activity is 0.2–2 µCi/µg protein. Actin labelled at higher specific activities is usually inactive.

Gel overlay assay

1. Separate duplicate samples by SDS-PAGE on mini-gels (0.5 mm thick). Fix the gel for 15 min in 50 per cent methanol, 10 per cent acetic acid.

Protocol 8. *Continued*

2. Remove the SDS and renature the protein by soaking the gel in several changes of 10 per cent ethanol for 1 h at room temperature.

3. Equilibrate the gel in 1 mM CaCl$_2$ in gelatin–PBS or 1 mM EGTA in gelatin–PBS for 1 h at room temperature.

4. Add ^{125}I-actin to the gel at a concentration of 5 µCi/ml and incubate for 5 h at room temperature or 12 h at 4 °C.

5. Remove unbound protein by washing the gel in 2–3 changes of calcium or EGTA–gelatin–PBS buffer until the counts in the wash solutions reach a minimum (usually 4–6 h at room temperature).

6. Visualize protein in the gel by staining with Coomassie Blue. Destain the gels and dry following normal procedures.

7. Detect the bound actin by autoradiography at −70 °C overnight. Usually 100 000 counts/min on the gel (detected by gamma counting) are required for an overnight exposure.

Notes

1. After dilution in the gelatin–PBS buffer, the actin is depolymerized to monomers. As a result, the assay will identify high-affinity interactions and not detect proteins that bind only to F-actin. Calcium-dependent actin-binding sites are detected by ^{125}I-actin binding to gel slices incubated in calcium but not in EGTA.

2. The assay is optimized for thin gels. Standard 0.75–1 mm thick gels require longer wash and incubation steps.

3. The overlay solution can be reused, but at the expense of removing the tightest-binding actin.

An overlay assay is powerful because it eliminates the need to purify and assay individual fragments. The fragments need only to be separated and resolved. SDS-PAGE is suitable for this purpose. When coupled with a cleavage map, the location of an actin-binding region can be determined. The resolution is limited by the complexity of the cleavage pattern. It is advisable to generate simple cleavage patterns because a small number of peptide fragments will be produced. Simple cleavage patterns are generated by cleaving at rare amino acids like cysteine. A calcium-dependent actin-binding site was mapped within the N-terminal 32 kDa of villin (5) using this approach. This location for a binding site is consistent with studies on fragments generated with proteases (*Figure 6b*).

Overlay assays are not suitable for finding binding sites for proteins that do not renature in SDS gels.

5. Conclusions and future directions

Positional information is a key aspect of protein structure that is usually lost in any peptide mapping study. As a result, it is relatively easy to detect and purify active fragments or domains of a protein, but it is more difficult to assign the activity to a region within the protein sequence. N-terminal sequence analysis and 'end'-specific polyclonal or monoclonal antibodies provide positional information and add a new dimension to peptide mapping studies. These methods form the basis of a scheme to map protein binding sites by a protein 'footprinting' approach (deArruda et al., in preparation). Because cleavage sites are easily identified with the antibody mapping approach, it is possible to extend the structural analysis of a protein by mapping regions that lie on the protein surface using site-specific chemical modification methods that in turn inhibit cleavage at the modified site. Surface sites that lie at a binding interface with a target protein are protected from chemical modification and are easily identified because those sites are now accessible to chemical or proteolytic cleavage.

Acknowledgements

I would like to thank my colleagues at the MRC Laboratory of Molecular Biology who helped me formulate the initial steps in domain mapping, and my colleagues at the Whitehead Institute and MIT for valuable discussions about mapping binding sites. The work described was supported by NIH DK-35306 and CA-47013, the Whitaker Health Sciences Foundation, and the Markey Foundation. During much of the work I was funded by a PEW Scholarship in the Biomedical Sciences.

References

1. Vandekerchove, J. (1990). *Curr. Opinions Cell Biol.*, **2**, 41.
2. Korn, E. D. and Hammer, J. A. (1990). *Curr. Opinions Cell Biol.*, **2**, 57.
3. Matsudaira, P. (1991). *Trends in Biochem. Sci.*, **16**, 87.
4. Matsudaira, P. T., Jakes, R., and Walker, J. E. (1985). *Nature*, **315**, 248.
5. Matsudaira, P., Jakes, R., Cameron, L., and Atherton, E. (1985). *Proc. Natl. Acad. Sci. USA*, **82**, 6788.
6. Matsudaira, P. (1987). *J. Biol. Chem.*, **262**, 10034.
7. Bazari, W. L., Matsudaira, P., Wallek, M., Smeal, T., Jakes, R., and Ahmed, Y. (1988). *Proc. Natl. Acad. Sci. USA.*, **85**, 4986.
8. Price, N. C. and Johnson, C. M. (1989). In *Proteolytic enzymes: a practical approach*. Beymon, R. J. and Bond, J. S. (eds.), p. 163. IRL, Oxford.
9. Bretscher, A. and Weber, K. (1980). *Cell*, **20**, 839.
10. Hesterberg, L. and Weber, K. (1983). *J. Biol. Chem.*, **258**, 359.

11. Hesterberg, L. and Weber, K. (1983). *J. Biol. Chem.*, **258**, 365.
12. Ishii, S. and Kumazaki, T. (1989). In *Methods in protein sequence analysis*, Wittmann-Liebold, B. (ed.), p. 156. Springer-Verlag, Berlin.
13. Janmey, P. A. and Matsudaira, P. T. (1988). *J. Biol. Chem.*, **263**, 16738.
14. Riveiere, L. R., Fleming, M., Elicone, C., and Tempst, P. (1991). In *Techniques in protein chemistry II*. Villafanca, J. J. (ed.), p. 171. Academic Press, San Diego.
15. Kabsch, W., Mannherz, H. G., Suck, D., Pai, E. F., and Holmes, K. C. (1990). *Nature*, **347**, 37.
16. Holmes, K. C., Popp, D., Gebhard, W., and Kabsch, W. (1990). *Nature*, **347**, 44.
17. DeRosier, D. J. and Tilney, L. G. (1982). *Cold Spring Harbor Symp. Quant. Biol.*, **66**, 525.
18. Joppich-Kuhn, R., Corkill, J. A., and Giese, R. W. (1982). *Anal. Biochem.*, **119**, 73.
19. Soutar, A. K. and Wade, D. P. (1989). In *Protein function: a practical approach*. Creighton, T. E. (ed.), p. 55. IRL, Oxford.
20. Snabes, M. C., Boyd, A. E., and Bryan, J. (1981). *J. Cell Biol.*, **90**, 809.
21. Matsudaira, P. T. and Burgess, D. R. (1978). *Anal. Biochem.*, **87**, 386.
22. Laemmli, U. K. (1970). *Nature*, **277**, 680.
23. Schagger, H. and von Jagow, G. (1987). *Anal. Biochem.*, **166**, 368.
24. Lunn, G. and Sansone, E. B. (1984). *Anal. Biochem.*, **147**, 245.
25. Spudich, J. A. and Watt, S. (1971). *J. Biol. Chem.*, **246**, 4866.
26. Pardee, J. D. and Spudich, J. A. (1982). In *Methods in cell biology*. Wilson, L. (ed.), Vol. 24, p. 271. Academic Press, New York.

5

The generation and isolation of mutant actins

R. KIMBERLEY COOK and PETER A. RUBENSTEIN

1. Introduction

A large number of studies have focused on the role of actin-based microfilament structures in processes such as muscle contraction, cytokinesis, cell-shape determination, cell motility, endocytosis, exocytosis, signal trans- duction, and cell polarization. These actin-dependent functions are regulated by a number of actin-binding proteins which control actin filament length, actin deposition in the cell, actin filament stability, and the ratio of G-actin to F-actin (1). Investigators have recently turned their attention to the molecular basis for the interaction between actin and the proteins which are thought to control its function *in vivo*. A recent study (2) using NMR has shown that troponin-I interacts with the N-terminal region of actin. Moreover, chemical cross-linking experiments have provided important clues concerning the sites of interaction of actin with various actin-binding proteins (3–6). Such studies, though, can only identify interaction sites where the chemistry of the cross- linker is compatible with active target residues on each of the proteins in question.

One approach for investigating protein–protein interactions is to use mutagenesis to make alterations in the primary structure of a particular protein and to assay the effect of the mutation on a protein function, either *in vivo* or *in vitro*. For such studies, one can make mutations ranging from deletions over a large range of residues to point mutations in individual residues. Furthermore, mutations can be introduced in a random fashion throughout different parts of the protein or directed to a specific site. Two major problems with mutagenesis studies are knowing where to make the mutations and expressing the mutant protein in a native form for functional studies.

The use of mutagenesis can potentially yield significant insights into actin biochemistry and the roles played by actin in cytoskeletal function and contractile processes. The genes for a number of different actins have been sequenced, and cDNAs have been isolated which makes mutagenesis possible

(7–17). Furthermore, the recent publication of the high-resolution three-dimensional structure of the actin-DNAse I complex allows an investigator to predict more reliably which regions of the actin molecule to mutate in order to study polymerization, cation binding, nucleotide binding, and the interaction of actin with various actin-binding proteins (18–19). New methods have also recently been developed for the preparation of quantities of mutant actins sufficient for biochemical studies of the protein *in vitro*, as well as for examination of the phenotypes associated with particular actin mutations *in vivo*.

Some of the work described in this chapter centres on the use of the yeast *Saccharomyces cerevisiae* as an experimental system. This organism has been valuable not only for studies of actin function but also for establishing the roles of a number of actin-binding proteins and tubulins in the biology of yeast (20). Cytoskeletal studies in yeast should provide valuable insight into the function of the cytoskeleton in higher eukaryotes because of the similarity of the proteins in these different cell types. General properties of yeast as an experimental system are described in Chapter 9. In this chapter we concentrate on the use of this system specifically for studying actin function.

2. Mutagenesis

The first stage in studying the effects of various mutations on actin function is to generate mutations in the coding sequence of an actin cDNA. For this task, one can use any of a large number of procedures designed to generate different types of mutations in different ways. Kits utilizing many of these procedures are commercially available, so the theory behind these methods is described here only in general terms (for review, see 21). Many of these procedures involve the use of a single-stranded DNA template containing the coding sequence of the protein involved.

One of the most popular routes for obtaining such a template is to clone the cDNA into a single-strand DNA bacteriophage such as M13mp18. A double-strand cDNA is cloned into the double-strand replicative form (RF) of the virus, which is then used to transform an F$^+$ strain of *E. coli* such as JM101 or JM105. Virus is then isolated from the infected culture and the single-stranded virus DNA prepared according to standard techniques. These M13 systems are commercially available from a number of sources.

Two general approaches to the generation of mutants can be employed, depending on the aim of the study. One can either introduce mutations randomly over the entire coding sequence of the actin cDNA, or direct them to a specific region of the DNA. This latter approach is termed site-directed mutagenesis.

2.1 Random mutagenesis

For the random approach, a convenient method is the use of bisulphite mutagenesis. In this approach, DNA is treated with sodium bisulphite, converting C → T, thus introducing point mutations randomly across the DNA (22). The mutated DNA is expressed in some system in which actin is required for viability, such as yeast (39), and one then selects colonies showing altered phenotypes under selective conditions such as abnormally high or low growth temperatures. In effect, natural processes are used to select mutations of interest for the investigator. The mutated actin coding sequence must then be isolated and sequenced to establish the site and nature of the mutation. The observed phenotype *in vivo* can be correlated with the mutational site, and one can attempt to isolate the mutant actin and study it *in vitro* as well as to try to establish the basis of the observed phenotype at a more defined molecular level. The advantage of such an approach is that one can potentially obtain a large number of mutant actins rapidly at little cost. Theoretically, these mutations can cover many areas of the actin polypeptide, giving rise to many colonies, each with a distinct phenotype. The disadvantages, however, are:

- An efficient selection procedure is needed to eliminate wild-type actin colonies.

- Many clones must be picked and the mutated DNA sequenced to find a selection of mutants.

- Not all possible mutations will be picked up under selective screens that might be used.

Furthermore, mutations in particular areas of the molecule of interest to the investigator may not be obtained, and many duplicate mutations in a few areas of the actin molecule may be isolated.

2.2 Site-directed mutagenesis

Currently, the most widely used protocols begin with single-stranded M13 phage DNA containing the actin coding sequence, as described above (23). One first synthesizes an oligodeoxynucleotide containing a mismatched base or bases at the site at which the mutation is to be introduced, placing the mismatched base in the middle of the oligonucleotide. If a simple point mutation is desired, an oligonucleotide 13–15 bases long is sufficient for the mutagenesis. If one wishes to alter a codon by two or three base changes or delete or insert codons, a longer oligonucleotide of 25–30 bases is needed.

Once the oligonucleotide is obtained, it is enzymatically phosphorylated and annealed to the single-stranded template DNA. The Klenow fragment of DNA polymerase is used to complete the complementary strand containing the mutated site, and the strand is circularized using T4 DNA ligase. Bacteria are transformed with the viral DNA, and plaques are examined for the presence of the mutated DNA using plaque-hybridization assays with the ^{32}P-labelled mutant oligonucleotide as the probe under restrictive hybridization conditions such that the probe does not anneal to the wild-type sequence. With this method, however, the number of mutant colonies compared to wild-type colonies observed is small. Therefore, two efficient and easy-to-use procedures have recently been developed to enhance greatly the mutant/wild-type ratio of viral plaques.

2.2.1 Thionucleotide method

The first procedure is based on the use of thionucleotides (24). The procedure outlined above is performed, except that synthesis of the complementary strand containing the mutated site is carried out in the presence of an α-thiophosphoronucleoside triphosphate. Only the mutant complementary strand will contain this modified nucleotide. The double-strand DNA is then treated with the restriction nuclease *Nci I*, which does not recognize thionucleotide sequences and cleaves only the parental wild-type strand. Exonuclease 3 is used to create large gaps which are filled in by Klenow fragment, copying the mutated sequence onto the new strand. When this DNA is used for transformation and plaques are isolated, up to 70 per cent carry the mutated actin DNA sequence. The plaque-hybridization previously required to identify the rare mutant sequences can thus be eliminated, and the phage DNA can be analysed directly by sequencing to identify the mutant DNA-containing clones and simultaneously establish that no extraneous mutations have been introduced. A commercial kit for using this procedure is sold by Amersham.

2.2.2 dUTP incorporation method (25)

Two enzymes in *E. coli* prevent the incorporation of dUTP into DNA. The first is a dUTPase encoded by the DUT gene, and the second is a uridyl N-glycosidase which hydrolyses uridine from DNA in which it has been incorporated. This glycosidase is encoded by the UNG gene. DNA cleavage takes place at this apyrimidine position, and excision repair mends the DNA through use of the complementary strand as a template.

For mutagenesis, the M13 containing the actin cDNA is grown in a bacterial strain that is dut⁻ and ung⁻. The resulting phage DNA contains a significant amount of dU in place of T. The mutant oligonucleotide is thus annealed to the dU-containing template strand as above, the complementary strand is completed in the absence of dUTP, and a dut⁺, ung⁺ strain is infected with the heteroduplex DNA. The bacteria selectively digest the original dU-containing wild-type strand, thereby greatly increasing the number of plaques containing mutant actin cDNAs. Since the frequency of the desired mutant plaques approaches 50 per cent with this method, the appropriate plaque can be selected by purifying and sequencing the viral DNA from 5–10 clones. BioRad Laboratories make a mutagenesis kit incorporating this methodology.

2.3 Preparation of an expression vector containing the mutant actin cDNA

Once a viral plaque containing the desired mutant actin cDNA has been identified and the integrity of the sequence confirmed, *E. coli* are infected with the virus containing the mutant actin cDNA, and a double-stranded RF is isolated from infected mass cultures of bacteria. Following isolation of the RF-DNA, a restriction fragment containing the altered DNA sequence is isolated. It is then subcloned into the appropriate expression vector carrying the wild-type actin cDNA with the fragment corresponding to the mutant fragment already removed. Once subcloning is accomplished, the expression vector carrying the mutated actin cDNA can now be introduced into the appropriate host cell by any of a number of different standard protocols. Detailed procedures for preparation of the RF form of M13 are included with both kits (Section 2.2) and are described in detail in numerous other places.

2.4 Recombinant circle polymerase chain reaction (26)

A method has recently been described, using the polymerase chain reaction (PCR), for making double-stranded site-directed mutant DNA directly without using the M13 phage system. This procedure eliminates the need to generate single-strand mutant phage, isolate it, sequence it, use it to isolate RF, and subclone the RF form into the expression vector.

This procedure requires four oligonucleotide primers and wild-type double-stranded template cDNA in an appropriate vector. The strategy is shown in *Figure 1*. Two overlapping complementary oligonucleotides each containing the desired mismatched base are prepared, and each is then utilized as a primer in a separate PCR reaction. For each reaction, the second primer on the complementary strand begins with the base following the breakpoint of the mutant oligonucleotide. The second two primers are chosen so that the breakpoints for each of the two reactions are different. The result is the production of two double-stranded linear DNAs, each containing the site of mutation. When these molecules are denatured and reannealed, two unligated double-stranded circular DNAs form, each carrying the desired mutation. Following transformation with these molecules, bacteria ligate the circles and amplify the plasmid. This plasmid can be isolated and used directly for production of the mutant protein, providing the proper expression vector was used initially. Although the use of PCR is associated with the introduction of random sequence errors, it is claimed that no erroneous bases were introduced in a 4 kb plasmid using this protocol (26). The use of this procedure is limited by the size of the plasmid in which the actin coding sequence is located. Since most expression vectors used are larger then 4 kb, use of this mutagenesis system will probably require a miniplasmid containing the cDNA to be used for the mutagenesis, after which the mutant double-stranded cDNA is subcloned into the expression vector for further study. This protocol should still be faster than the M13 system which alternates between single-stranded and double-stranded DNA.

2.5 Simultaneous production of multiple mutations

It is often difficult to make a conclusion about structure/function relationships based on a single amino acid substitution. Rather, such conclusions should be based on the results obtained by a series of amino acid substitutions at the same site. To do this economically, one can synthesize the mutant oligonucleotide with different combinations of the four bases in single positions (27). Following mutagenesis with one of the methods outlined

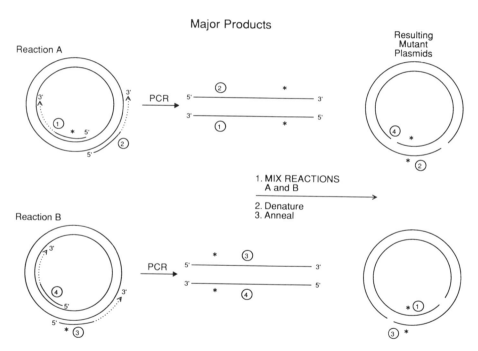

Figure 1. Recombinant circle mutagenesis. For reactions A and B, the double circle represents the plasmid carrying the actin coding sequence. The asterisk represents the site of the mutation. The solid lines 1–4 represent the four primer oligodeoxynucleotides, as described in the text, and the dotted lines represent the action of the DNA polymerase in extending these primers to make complete circles. Note that for each pair of primers used in reaction A and B, only one oligodeoxynucleotide carries the mutation. The 'breakpoint' described in the text is the site in the plasmid where a break is introduced in making the two members of the oligodeoxynucleotide pair

above, a number of clones can be sequenced to determine which ones exhibit non-identical base changes and to establish the sequence of the particular clone. In this manner, multiple mutations at a single site can be obtained for the cost of a single oligonucleotide.

An interesting extension of the combined random site-directed mutagenesis approach in the study of actin structure/function relationships has been utilized recently by Ken Wertman, David Botstein, and David Drubin (personal communication). These investigators reasoned that since charged residues were likely to be found on the surface of the protein, they would be the most likely to be involved in interactions between actin and actin-binding proteins. To identify these important residues, they used a technique called alanine scanning mutagenesis, first applied to the study of the hGH-receptor (28). In this method they individually changed every Asp and Glu residue to an Ala residue, using a series of 36 oligonucleotides carrying site-directed base changes. Basic residues, also likely to be on the surface of the protein, were not addressed in this initial study. The resulting actin mutants were expressed in yeast, and the actins from transformed cells with aberrant phenotypes were selected for further study. This array of mutants also identified domains on the three-dimensional structure of actin which were most likely involved in protein–protein interactions.

3. Expression of mutant actins

3.1 An overview of potential systems

The most difficult part of applying the use of site-directed mutagenesis to the study of actin structure/function relationships is the expression of the protein in a native form for subsequent study. Until recently, two systems were generally used by investigators to overexpress mutant proteins: yeast and bacteria. *Table 1* summarizes the advantages and disadvantages of the systems discussed in this paper.

3.1.1 Expression of actin in bacteria

A number of investigators succeeded in producing actin in bacteria following transformation with a plasmid carrying an actin cDNA. However, when normal cell disruption methods were employed, the actin produced by the bacteria was isolated as an insoluble inclusion body which could not be successfully unfolded and renatured to yield native actin (29).

This folding problem appears to have been successfully solved by the approach of Frankel *et al.* (ref. 30; described in Sections 3.3 and 4.2) which uses a lysis procedure based on the detergent Sarkosyl, a development which may make the large-scale production of mutant actins feasible. Two problems still remain with this system, however. The *Dictyostelium discoideum* actin cDNA used has a very strong internal translation initiation site, such that

Table 1. Comparison of mutant actin expression systems

System	Advantages	Disadvantages
in vitro	high-specific-activity labelling synthesis of dominant lethal mutant actins possible	small amounts synthesized some of the actin may be non-native
bacteria	large quantities possible synthesis of dominant lethal mutants in eukaryotes possible	poor yields unmodified, or incorrectly modified
Drosophila flight muscle	genetic manipulatability well-ordered structures for studying morphological effects of a mutant actin on muscle structure	hard to purify complex experimental system to establish difficult to correlate morphology with disruption of specific protein–protein associations
yeast	genetic manipulatability correlation of *in vivo* and *in vitro* effects production of large quantities of mutant actins possible	small amount of actin per cell few assayable actin functions *in vivo* partial or completely dominant actins hard to obtain in pure form

more than 90 per cent of translation begins one-quarter of the way into the coding sequence, yielding a truncated protein. Therefore, only a small amount of the actin produced in a transformed culture will be amenable to further study. This amount is additionally reduced by the purification scheme discussed below (Section 4.2). If one could, through mutagenesis, inactivate this internal start site, the usefulness of this system should be much improved.

The second potential problem with the actin produced in bacteria centres on the post-translational modifications that actin undergoes in eukaryotic cells. With one exception (31), all actins studied to date have a 3-methylhistidine residue at His_{73} (32–34). Bacteria probably do not possess an actin methylase since they normally do not synthesize actin. We have shown by four assays of actin function *in vitro* that the absence of this methyl group has no adverse effects on actin behaviour (35). Furthermore, D. Shortle and co-workers (personal communication) have demonstrated through a muta-genesis study that the absence of this methylhistidine in yeast actin has no effect on yeast viability. Nevertheless, the conservation of this modification throughout the eukaryotes strongly suggests that it may be functionally important in some context. The results using bacterial actin might therefore be misleading if the actin function being addressed involves the methylhistidine residue.

A second common post-translational modification of actin occurs at the N-terminus of the primary translation product. For most actins, the genes code for one or two more amino acids than are found in the mature molecule, which usually begins with an Ac–Asp or Ac–Glu. These extra amino acids, usually Met and Cys, are first acetylated and then removed as acetylated amino acids by a specific actin N-terminal-processing enzyme following release of the actin from the ribosome. The unblocked amino terminus of the resulting polypeptide is then reacetylated (36–37). The functional significance of this processing event remains unknown. However, a series of studies has shown that it occurs very near a site through which actin interacts with a number of actin-binding proteins (2–6).

Bacteria do not carry out this processing event, and they certainly do not acetylate proteins on their N-termini. In bacteria, the N-terminus of proteins may have one of three structures. They may begin with an N-terminal formyl-Met, the residue with which translation is initiated. Alternatively, either the formyl group alone or both the formyl and Met residues can be removed. If the N-terminal structure of actin is crucial for proper actin function, some results obtained with bacterially synthesized actin could be misleading.

3.1.2 Expression in yeast

Initial attempts to use yeast as an expression system for actin were unsuccessful. Although actin is essential for yeast viability, overexpression of actin in yeast is lethal (38). We and others have recently overcome this problem by controlling the level at which the mutant yeast actin is expressed. Although this manipulation can be handled genetically in a number of ways (39–40), we outline here only the procedure which we currently utilize. We begin with a diploid strain of Δleu2, ura3-52 yeast in which one of the two actin genes has been inactivated by insertion of a wild-type LEU2 gene, making the cells LEU$^+$ (act1::LEU2/ACT1$^+$). The cells are then transformed by the plasmid yCp50 carrying a mutant yeast actin cDNA adjacent to the yeast actin promoter. This plasmid carries an ARS1 sequence, allowing it to be autonomously replicated. It also contains a CEN4 sequence which limits the plasmid copy number and works to ensure segregation of the plasmid into both the mother and daughter cells during budding. The plasmid also carries a URA3$^+$ gene. Thus, one isolates transformed diploids by selecting URA$^+$, LEU$^+$ cells, guaranteeing that the selected cells have one active wild-type and one active mutant actin gene. Once transformed diploid cells are obtained, the effect of the mutant actin on the yeast can be studied *in vivo*.

By altering the culture conditions one can induce the yeast to undergo meiosis. This process yields four haploid spores. If segregation of the plasmid is complete, then two of these (URA$^+$, LEU$^+$) contain only the mutant actin gene as the functional gene and two contain both the mutant and the wild-type actin genes (URA$^+$, leu$^-$). If URA$^+$, LEU$^+$ haploid cells are obtained, one can conclude that the mutant actin is capable *by itself* of supporting yeast life and can then study the phenotype of the cell to determine if there is a specific defect caused by the mutation. One can then also grow the mutant-carrying yeast in large culture and prepare the homogeneous mutant actin for biochemical studies. If only the transformed diploid cells are viable, one must try to separate the mixture of mutant and wild-type yeast actins in order to study the mutant protein biochemically.

Two general problems are associated with this system. First, actin in yeast is present at less than one-tenth the level found in higher eukaryotic cells, making the purification much more difficult. This problem is especially troublesome if affinity chromatography procedures will not work with the mutant actins. Second, we have recently discovered that *S. cerevisiae* and three other fungi, unlike other eukaryotes, do not process actin at the N-terminus although, under normal circumstances, they do produce acetylated actin (61). Thus, caution must be exercised in applying results from structure/function studies of yeast actin to the behaviour of actins in higher eukaryotes due to the presence of an extra N-terminal amino acid.

3.1.3 Expression in a cell-free translation system

A third approach to expression of actin is to synthesize the protein *in vitro* using a cell-free translation system. In this procedure, one clones the actin cDNA into a plasmid such as one of the pGEM vectors (Promega Biotech) so that the coding sequence is adjacent to either a bacteriophage T7 or SP6 promoter. One can then use this plasmid together with either T7 or SP6 RNA polymerase, both commercially available, in the presence of 7-methyl-GpppG to synthesize 5' capped actin mRNA *in vitro* (41). This mRNA can be utilized as a template in a rabbit reticulocyte cell-free translation system containing a radioactively labelled amino acid to make small quantities of highly labelled actin. Lysate preparations are available as complete kits from a number of commercial sources. The labelled actin can then be used for binding studies with other actin-binding proteins or for studies on various aspects of actin metabolism. The procedure is useful if a particular mutant actin is lethal for yeast or bacteria.

There can be problems with this method, however. First, the commercially available system is expensive to use in large amounts. Second, one cannot isolate enough of the translation product to study aspects of actin behaviour such as metal or nucleotide binding or the activation of myosin ATPase activity. Third, in our hands as well as in the hands of others, the mRNAs of different actin isoforms appear to be translated with different efficiencies in different preparations of reticulocyte lysate. These differences can be as much as three- or fourfold in magnitude. However, since actin has such a highly conserved primary structure among organisms and tissues, it is often possible to substitute one actin for another for a particular study in order to maximize the efficiency of the reticulocyte lysate system. Finally, there are several indications that the actin produced *in vitro* may not be in a fully native conformation (unpublished observations). For example, its ability to bind completely to DNAse-agarose and its ability to be processed completely at its

N-terminus may be changed. To overcome this problem, we perform parallel translations of wild-type and mutant actins, each labelled with a different radioisotope, such as ^{3}H-leucine and ^{35}S-methionine, and mix the two actins together. We then determine the isotope ratio of the mixture before and after a particular experiment analysing an aspect of actin function. A difference in this ratio would indicate a functional difference brought about by the particular mutation (42).

3.1.4 Expression of mutant actin in a flight muscle actin-negative, flightless *Drosphila melanogaster*

A fourth expression system makes use of the *Drosophila* indirect flight muscle, and has been especially exploited by Fyrberg and co-workers (43–45). The only actin made in this tissue is encoded by the act88F gene; flies with this gene inactivated have been isolated, as these flies live but cannot fly. Using this Act88f$^-$ fly as a host, one introduces an exogenous wild-type or mutant actin gene into the germ cells via P-element mediated transformation and examines the ability of the offspring expressing the exogenous actin to recover flight. Even if flight is not restored, one can then carry out an electron microscopic examination of the flight muscle tissue to judge the effect of the mutation on the organization and assembly of sarcomeric structures. A number of these mutations have been examined and a wide range of subcellular phenotypes have been observed. The strength of this method is that one can correlate a particular mutation with a specific irregularity in a well-characterized and normally well-organized subcellular structure. The problems with the system are threefold. First, it is difficult to associate a particular type of disorganization with a specific disruption of an interaction between actin and a certain actin binding protein. Second, it has so far not been feasible to extract the mutant actins in native form from the flies in amounts needed for extensive investigation of the mutant actins *in vitro*. Third, this is not a facile experimental system except for experienced *Drosophila* biologists. We will not give detailed protocols for these procedures here. They are, however, described in detail in the papers by Fyrberg's group (43–45).

3.2 Labelled actin synthesized in a cell-free translation system (42)

For this method, one begins with an actin cDNA adjacent to a bacteriophage SP6 promoter in a pGEM expression vector, linearizes the plasmid, synthesizes RNA *in vitro*, and uses this RNA to direct actin synthesis *in vitro*. The procedure for making linear DNA is given in *Protocol 1*, and preparation of actin mRNA is given in *Protocol 2*.

Protocol 1. Preparation of linear DNA

1. In a tube mix 100 µg closed circular actin plasmid DNA, 30 µl of 10× *Eco RI* buffer (500 mM Tris-HCl, pH 8.0, 100 mM MgCl$_2$, and 1 M NaCl) and water, untreated with DEPC, to 285 µl.

2. Add 15 µl *Eco RI* (10 000 U/ml). Incubate for 2 h at 37 °C.

3. Extract twice with an equal volume of phenol, retaining the RNA in the aqueous layer.

4. Extract twice with an equal volume of chloroform and keep the upper layer.

5. Add 30 µl of 3 M sodium acetate; then add 900 µl of 95 per cent ethanol. Place the solution in dry ice for 15 min, and collect the precipitate by centrifugation at 15 000 × *g* at 4 °C for 15 min.

6. Wash the pellet with 70 per cent ethanol and dry by lyophilization.

7. Dissolve the pellet in 100 µl of DEPC-treated water to give a final concentration of approximately 1 mg of linear DNA/ml.

Protocol 2. Generation of actin mRNA *in vitro*

1. Make a 5× transcription buffer solution consisting of 2 M Tris-HCl, pH 7.5, 30 mM MgCl$_2$, and 10 mM spermidine. Use DEPC-treated water for the synthesis of the RNA.

2. Bring the following components to room temperature and then combine in order:

   - 190 µl water
   - 100 µl transcription buffer
   - 50 µl 100 mM dithiothreitol
   - 15 µl RNAsin at 40 U/µl (Promega)
   - 100 µl of a mixture of NTPs (2.5 mM each of ATP, CTP, UTP, and 2 mM GTP)
   - 25 µl of 7-^{me}G-pppG (10 mM)
   - 15 µl of linearized actin DNA (1 µg/ul)
   - 5 µl (100 units) of SP6 DNA-dependent RNA polymerase

3. Incubate the reaction for 1.5 h at 37 °C.

4. Add 5 µl RNAse-free DNase to remove the DNA, mix and incubate for 15 min at 37 °C.

5. Extract twice with phenol and once with chloroform, keeping the upper aqueous phase at each step; add 0.1 volume of 3 M sodium acetate and 3

volumes of 95 per cent ethanol to precipitate the RNA. Collect the RNA by centrifugation at 15 000 × g for 15 min at 4 °C.

6. Wash the resulting precipitate with 70 per cent ethanol, lyophilize, and resuspend the pellet in 100 μl untreated, double-autoclaved water. This preparation provides enough RNA for 20 translation reactions. This water should *not* be treated with DEPC, since residual amounts of ethanol produced by the degradation of DEPC will severely inhibit translation of the mRNA in the next step.

3.2.1 Synthesis of radioactive actin

The reaction utilizes a rabbit reticulocyte cell-free translation system commercially available from a number of vendors. For this reaction, micropipette tips should be autoclaved and baked at 100 °C overnight. Water should be double-autoclaved but not DEPC-treated (see *Protocol 2*). The amino acid mixtures used for the reaction contain 19 of the 20 amino acids needed for protein synthesis. The one omitted is replaced by the labelled amino acid of choice. Kits can be routinely purchased to allow labelling with methionine, cysteine, or leucine. If other labelled amino acids are to be used, the unlabelled amino acid mixtures should be composed of 2 mM of each of the remaining 19 amino acids.

For the translation reaction, combine 35 μl of micrococcal nuclease-treated reticulocyte lysate, 1 μl unlabelled amino acid mix, 5 μl radioactive amino acid solution, 1 μl RNAsin, 4 μl RNA solution from *Protocol 2*, and 4 μl water. Micrococcal nuclease treatment destroys the endogenous globin mRNA which would otherwise overwhelm the translation system. Incubate the mixture for 1 h at 30 °C. When [^{35}S]-methionine (> 1000 Ci/mmol) is used, this protocol should yield about 100 000–150 000 dpm of labelled actin/ 2 μl of translation mixture.

3.2.2 Additional comments

Actin mRNA in general is not as good a template for translation as the bromomosaic virus RNA provided as a standard in most of the translation kits available. Additionally, translation of actin mRNA is much more sensitive to various inhibitors than is translation of bromomosaic virus RNA. Finally, the RNAs for different actins will behave differently from one another in terms of efficiency of translation in a single lysate preparation, and different lysate preparations will have varying efficiencies for translating different actins. Thus, for maximum yield of labelled actins, one has to establish the best conditions overall for a particular actin mRNA.

We have also found up to a three- or fourfold difference in the efficiency of translation using different [^{35}S]-methionine preparations of nearly equivalent specific activities from different companies. Some of these preparations come with chemical stabilizers which can, in certain instances, suppress translation levels. The best combination of lysates and methionine must be experimentally determined.

Some labelled amino acids come in an aqueous ethanol solution. Since ethanol inhibits translation reactions, the ethanol must be removed from these preparations by lyophilization before use.

3.3 Actin synthesized in *E. coli* (30)

An actin cDNA is cloned into an inducible bacterial expression plasmid such as pKK233 (Pharmacia) which carries a *Tac* promoter that responds to either tryptophan or isopropyl thiogalactoside added to the medium. Care must be taken to eliminate all nucleotides of the cDNA upstream from the initiator ATG codon before cloning the actin coding sequence into the vector. Otherwise, the additional nucleotides will interrupt the Shine–Dalgarno sequence which directs ribosome binding to the message at the translation initiation site (46). This trimming can be carried out by the use of specific endonucleases or by the exonuclease *Bal 31*, and requires an additional input of effort. The bacteria are transformed with the plasmid, and transformants are selected for the appropriate antibiotic resistance. *Protocol 3* is based on the work of Frankel *et al.* (30).

Protocol 3. Growth of bacterial cultures for actin expression

1. Grow small cultures of transformed bacteria overnight at 37 °C in LB (1 per cent tryptone, 0.5 per cent yeast extract, 0.5 per cent NaCl).

2. Place 10 ml of overnight culture and 1 l of M10 medium (47) containing 40 μg/ml of ampicillin, 0.5 per cent Casamino acids, 1 per cent glycerol, 5 per cent LB broth, and twice the amount of ammonium chloride as in normal M10 medium in a 2 l baffled flask and shake at 37 °C.

3. When the cultures reach an OD_{550} of 0.2, add IPTG to a final concentration of 4 mM.

4. When the cells have reached an OD_{550} of 1.0, chill the cultures in ice water for 30 min and harvest the bacteria by centrifugation.

5. Purify the actin from these bacteria using the procedures described later.

3.4 Actin synthesized in *S. cerevisiae*

3.4.1 Transformation of yeast with mutant actin coding sequences

Once the mutant actin cDNA of interest has been generated, it must be subcloned into a vector that will regulate the expression and copy number of the actin sequence in yeast. A diploid yeast strain such as TDyDD (αΔleu2/ aΔleu2; ura3-52/ura3-52; act1::LEU2/ACT$^+$) carrying one disrupted actin gene is transformed with a centromeric plasmid carrying the URA3 gene such as yCp50. This plasmid may carry the wild type yeast actin sequence or a mutant sequence. *Protocol 4* is based on the method of Ito *et al.* (48).

Protocol 4. Transformation of yeast

1. Grow 40 ml of cells to $1-3 \times 10^7$/ml in YPD (2 per cent glucose, 1 per cent yeast extract, 1 per cent peptone). Harvest the cells by centrifugation. Wash twice with 10 ml sterile TE (10 mM Tris, pH 7.5, 2 mM EDTA).

2. Resuspend the cells in TE to a density of 2×10^8/ml.

3. Mix 0.5 ml of cells with 0.5 ml of 0.1 M LiCl in a 12 ml snap-cap tube. Shake at 30 °C for 1.5 h.

4. Aliquot the DNA for transformation into 1.5 ml Eppendorf tubes. Use no more than 20 μg DNA in $\leq$ 25 μl of solution. Include a negative control (25 μl TE) and positive controls which include yCp50 and pCEN-WT, a plasmid carrying the wild-type actin sequence.

5. Add 100 μl of the LiCl-treated cells to each tube of DNA. Let the cells incubate at 30 °C for 30 min.

6. Add an equal volume of 70 per cent PEG-3350 (this was formerly called PEG-4000) to each tube. Incubate at 30 °C for 1 h.

7. Heat shock the cells at 42 °C for 6 min.

8. Immediately cool the cells to 20 °C in water for 3 min.

9. Pellet the cells, remove supernatant, and wash with 1 ml of sterile water.

10. Pellet the cells again and resuspend in 0.5 ml water.

11. Plate 250 μl, 100 μl, and 10 μl of the transformed cells out on selective plates (lacking uracil, in this case). To determine the survivability of the cells and the transformation efficiency, dilute the cells 1:1000 and plate 100 μl of this on a YPD plate.

12. Incubate 1.5–2 days at 30 °C and check for URA$^+$ colonies. If no colonies appear on the plates for the mutant actin, incubate the plates for several days more. The cells may grow slowly if the actin mutation is semi-dominant. The mutant actin could also be completely dominant, leading to the death of transformed cells.

3.4.2 Isolation of haploid cells producing only the mutant actin

Once diploid URA$^+$ transformants have been isolated, one can induce these cells to undergo meiosis and sporulation and try to isolate URA$^+$ LEU$^+$ haploids (*Protocol 5*).

113

Protocol 5. Isolation of haploid cells producing mutant actins

1. Replica plate the cells on to a sporulation plate containing 2 per cent potassium acetate, 0.1 per cent glucose, 0.25 per cent yeast extract supplemented with 20 amino acids at 0.001 per cent each. Incubate 3–5 days at 30 °C. Alternatively, transfer a growing culture of cells to liquid sporulation medium (the same as described above) and shake for several days at 30 °C. Check the extent of sporulation by microscopic examination.

2. Digest the spores with 0.5 mg/ml Zymolyase 100T (ICN Biochemicals), in 1 M sorbitol for 30 min at 30 °C. This procedure digests the ascus of the spore, allowing the cells to be separated.

3. Spread this suspension out on a plate and separate the four spores of each tetrad with a micromanipulator.

If a micromanipulator is not available, haploid URA$^+$, LEU$^+$ cells can still be obtained if the initial strain of cells used is can^r/can^s (canavanine resistant or sensitive). Plate a culture of sporulated cells out on a plate containing canavanine (an arginine analogue) but lacking uracil and leucine. Colonies which grow will be haploid, due to the recessive nature of canavanine resistance, and carry only the mutant actin sequence.

Any haploid URA$^+$, LEU$^+$ cells isolated by either method will produce the mutant actin only and can be used for the production of mutant protein for biochemical studies.

3.4.3 Plasmid rescue from URA$^+$, LEU$^+$ cells

Regardless of the method of isolation of haploid cells, it is important to 'rescue' the plasmid from the yeast in order to determine that the mutation is still in place and that no other mutations have been spuriously introduced. This rescue procedure is described in *Protocol 6*.

Protocol 6. Plasmid rescue

1. Pellet the transformed yeast from a 1.5 ml overnight culture.

2. Resuspend the cells in 200 μl of a solution containing 1 per cent SDS, 2 per cent Triton X-100, 100 mM NaCl, 10 mM Tris, pH 8.0, 1 mM EDTA.

3. Add an equal volume of 0.5 mm glass beads. Then add 200 μl of phenol:chloroform:isoamyl alcohol (25:24:1) (v/v).

4. Vortex the mixture for 2 min at room temperature.

5. Centrifuge for 5 min in a microcentrifuge and retain the aqueous layer.

6. Re-extract the aqueous layer with another 200 μl of phenol:chloroform: isoamyl alcohol.

7. Ethanol-precipitate the yeast nucleic acids.

8. Resuspend the pellet in 400 μl TE. Add 5 μl of 10 mg/ml RNase A and incubate at 37 ° for 30 min.

9. Extract the DNA with phenol once and chloroform once, retaining the aqueous layer each time, and ethanol-precipitate.

10. Resuspend the pellet in 50 μl of TE. Make a series of dilutions (1:10, 1:100, 1:1000) of this DNA.

11. Transform competent *E. coli* with the different dilutions of DNA (techniques for transformation of *E. coli* are detailed in ref. 49). Ampicillin-resistant colonies will contain the same plasmid as found in the yeast.

   • **Note**: It is very important that the transformation efficiency of these bacteria be as high as possible. The yeast chromosomal DNA makes this procedure very inefficient.

12. Isolate the plasmid from the bacteria and sequence to reconfirm the mutation.

4. Mutant actin purification protocols

4.1 Radiolabelled actin from a cell-free translation reaction

Protocol 7 is based on the purification of actin from 200 μl of translation mixture. It will give nearly radiochemically pure actin, although the final solution is contaminated by unlabelled proteins from the reticulocyte lysate.

Protocol 7. Purification of actin from translation mixtures

1. Place the translation mixture containing the labelled actin over a 10 ml G-25 Sephadex column equilibrated in G-buffer (10 mM Tris-HCl, pH 7.5, containing 0.1 mM ATP, 0.1 mM MgCl$_2$, and 1 mM 2-mercaptoethanol) to which has been added 0.1 M KCl and 10 per cent (v/v) formamide.

2. Collect fractions of 250 μl each and pool red-coloured fractions.

3. Equilibrate a 1 ml DE-52 DEAE cellulose (Whatman) column with G-buffer plus 0.1 M KCl. Make sure that the effluent buffer has the same pH and ATP content as the original buffer.

4. Apply the red fractions from the G-25 column to the DEAE column.

5. Wash with G-buffer containing 0.1 M KCl and 10 per cent formamide.

6. Wash with G-buffer containing 0.15 M KCl and 10 per cent formamide.

7. Elute the actin with G-buffer containing 0.3 M KCl. Collect 200 μl fractions and pool the peak of radioactivity.

8. If necessary, transfer the actin into G-buffer by dialysis.

4.2 Actin synthesized in bacteria

4.2.1 Sarkosyl lysis

Protocol 8 is taken from Frankel *et al.* (30) and was developed using actin from *Dictyostelium discoideum* synthesized in bacteria.

Protocol 8. Lysis of bacteria in Sarkosyl. (All procedures should be done at 4 °C)

1. Wash the bacteria from 1 l of culture with 20 mM Tris-HCl, pH 8, containing 50 mM NaCl.

2. Suspend the cells in 15 ml STE (10 per cent sucrose, 100 mM Tris-HCl, pH 8, 1.5 mM EDTA). Add 1.5 mg of lysozyme and incubate 15 min.

3. Add 132 ml of lysis buffer (15 mM triethanolamine-HCl, pH 8, 50 mM NaCl, 2.5 mM ATP, 1 mM GDP, 1 mM dithiothreitol, 20 µg/ml aprotinin, 10 µg/ml leupeptin, 5 µg/ml pepstatin, 2.5 µg/ml chymostatin, 0.43 mM phenylmethylsulphonyl fluoride, and 0.43 mM O-phenanthroline).

4. Add 10 ml of Sarkosyl while stirring to give a final concentration of 0.2 per cent. Let sit for 2 min.

5. Sonicate the mixture, using seven 10 s bursts at 90 W, to reduce the viscosity of the solution.

6. Centrifuge for 11 min at 32 000 × g.

7. Add octylglucoside to the supernatant to a final concentration of 2 per cent. Stir for 5 min.

8. Add MgCl$_2$ and CaCl$_2$ to final concentrations of 1.25 and 1.0 mM, respectively, and stir for 20 min.

9. Centrifuge the solution for 12 h at 60 000 × g to pellet the 30 S ribosomal subunits from the actin-containing supernatant, and further purify the supernatant actin as described in *Protocols 9* and *10*.

4.2.2 DNase I-affinity chromatography (50)

To separate the actin from other *E. coli* proteins, DNase I-affinity chromatography is utilized as in *Protocols 9* and *10*.

Protocol 9. Preparation of DNase I-agarose

1. Wash one 25 ml bottle of Affigel 10 (BioRad) into a Buchner funnel with a sintered glass filter. Wash well with cold water to remove the isopropanol but do not allow the gel to dry.

2. Transfer the cake of gel to a 50 ml screw-cap tube. Add Hepes, pH 7.4, and $CaCl_2$ to make a final concentration (including gel volume) of 0.1 M and 2 mM respectively.

3. Add 100 mg of DNase I (Worthington Biochemical). Shake gently at 4 °C for 4–5 h.

4. Spin down the gel at 500 × g for 5 min. Remove the supernatant and add 0.5 M Tris, pH 7.5, to block any remaining active sites. Shake gently 4–5 h at 4 °C.

5. Spin down the gel at 500 × g for 5 min, remove the supernatant and wash the gel with 3 M guanidine-HCl containing 0.5 M sodium acetate.

6. Remove the guanidine solution and wash the gel well with 10 mM Tris, pH 7.5, 2 mM $CaCl_2$.

7. Store the DNase I-agarose at 4 °C in this solution containing 0.02 per cent NaN_3.

Protocol 10. DNase I-affinity chromatography to purify *E. coli*-synthesized actin

1. Mix the supernatant obtained by *Protocol 8* with 5–10 ml of packed DNase I-agarose.

2. Batch-wash the resin with 15 bed volumes of high-salt buffer (25 mM triethanolamine, pH 8.0, 0.8 M NaCl, 1 mM ATP, 1 mM GDP, 5 mM sodium pyrophosphate, 0.25 mM dithiothreitol, 0.1 mM EDTA, 1.4 mM $MgCl_2$, 1 mM $CaCl_2$, and 0.5 mM PMSF) and pour into an appropriately-sized syringe. Wash with another five volumes of high-salt buffer. The GDP keeps EF-Tu, a major contaminant, in a native form, allowing its separation from the actin.

3. Drain most of the buffer from the column and resuspend the resin in 0.3 column volumes of 100 per cent deionized formamide. Incubate for 10 min.

4. Completely drain the resin bed, forcing all liquid through by air pressure.

5. Dilute the eluate with high-salt buffer so that the final concentration of formamide is 30 per cent.

4.2.3 Further purification of actin

This is carried out by Sephadex G-150 chromatography (*Protocol 11*).

Protocol 11. Sephadex G-150 chromatography

1. Immediately clarify the eluate from the DNAse I column by centrifugation and load on to a 1.5×68 cm column of G-150 Sephadex equilibrated in 25 mM triethanolamine, pH 8.0, 0.8 M NaCl, 0.2 mM ATP, 5 mM sodium pyrophosphate, 0.25 mM dithiothreitol, 0.1 mM EDTA, and 1 mM $CaCl_2$.

2. Develop the column in the same buffer, and pool fractions containing the actin as determined by calibration of the column with muscle actin.

3. Concentrate the pooled actin by vacuum dialysis against a buffer containing 2.5 mM triethanolamine-HCl, pH 8.0, 0.5 mM dithiothreitol, 0.2 mM ATP, and 0.1 mM $CaCl_2$ to a volume of 250 μl. Polymerize the actin as described in Section 4.2.4.

4.2.4 Polymerization of the actin

To the concentrated actin solution from the previous step must be added ATP to 1 mM, $MgCl_2$ to 4 mM, NaCl to 50 mM, and phalloidin to 25 μM: incubate overnight at 0 °C. Phalloidin stabilizes the F-actin but will interfere with subsequent functional assays requiring G-actin. Collect the F-actin by centrifugation for 30 min in a Beckman TL-100 tabletop ultracetrifuge at $130\,000 \times g$.

4.3 Actin synthesized in yeast

4.3.1 Chromatographic procedure

The purification of yeast actin by a series of chromatographic techniques has been reported in the literature (51). However, we and others have had difficulty with this protocol and it will not be described here in greater detail. An apparent improvement in this approach has recently been reported in abstract form by Scordilis and Anable for actin from *Schizosaccharomyces pombe* (52). They begin with an acetone powder made from yeast spheroplasts generated by digestion of the yeast cell walls with a yeast lytic enzyme preparation in hyperosmotic medium. Extraction of the acetone powder with a depolymerizing buffer is followed by chromatography of the extract on ATP-saturated DEAE-Sepharose using a gradient from 0.1 to 5 M NaCl. Subsequent chromatography of the actin-containing fractions on a Sephacryl S-100 HR column produced a very highly purified actin preparation. Pure actin was generated by a cycle of polymerization–depolymerization (60), using 2 mM $MgCl_2$ to effect the polymerization. With this protocol, the actin isolated was about 1 per cent of the protein extracted from the acetone powder. This protocol should be especially useful for mutant actins which will not bind to a DNAse I-agarose affinity column.

4.3.2 Purification of actin using DNase I chromatography

This is a fast and simple procedure, slightly different from that used for purification of bacterial actin, that yields large quantities of native yeast actin. Its only drawback is that some mutant actins, those at Asp_{11} for example, do not bind well to DNase I. Details are given in *Protocol 12*.

Protocol 12. Purification of yeast actin using DNase I-affinity chromatography. This is based on the methods of Zechel (50) and Kron and Spudich (personal communication)

1. Grow 6 l of cells to at least $OD_{660} = 2.0$. Harvest the cells by centrifugation.

2. Resuspend the cells in G-buffer plus protease inhibitors (PIs): 10 mM Tris, pH 7.5, 0.2 mM $CaCl_2$, 0.2 mM ATP to which is added the following PIs (per litre of G-buffer): PMSF—1.2 ml of 250 mM in ethanol; aprotinin—500 μl of 1 mg/ml in water; antipain—500 μl of 1 mg/ml in water; leupeptin—500 μl of 1 mg/ml in water; pepstatin—500 μl of 1 mg/ml in DMSO; TPCK—500 μl of 1 mg/ml in ethanol.

3. Place the cells in the pre-chilled large chamber of a Beadbeater (Biospec Products) already filled half-way with glass beads (0.5 mm diameter) which have been wetted with G-buffer plus PIs. Fill the chamber to the top with more G-buffer plus PIs.

4. Lyse the cells by turning on the Bead Beater for 20 s. Allow 2 min rests between bursts to allow the lysate to cool. Repeat three times and check the cells for lysis by microscopy. Continue 'beating' until most of the cells have lysed. Remove the beads by passing the lysate through cheesecloth or gauze. Rinse the beads briefly with G-buffer plus PIs, adding this wash to the cell lysate.

5. Centrifuge the lysate at 100 000 × g for 45 min at 4 °C to clear it.

6. Remove as much lipid from the supernatant as possible and apply the supernatant to a 6 ml DNase I column in G-buffer. We use the BioRad 10 ml Dispo columns for this procedure.

7. Wash the column with 10 ml G-buffer containing PIs.

8. Wash the column with 10 ml G-buffer containing PI's and 0.4 M NH_4Cl.

9. Wash the column with 10 ml G-buffer containing PIs.

10. Elute the actin with 10 ml of G-buffer containing 50 per cent formamide. Drip the eluted protein solution directly on to a 200 μl DEAE DE-52 column which has been washed well in G-buffer.

11. Wash the DEAE column with 1 ml G-buffer.

Protocol 12. *Continued*

12. Elute with G-buffer containing 0.3 M KCl, collecting 100–200 μl fractions. Pool the peak protein fractions.

13. Polymerize the actin by addition of $MgCl_2$ to 2 mM. This is necessary because some of the actin (< 10 per cent) is denatured during this procedure. This protocol will yield 1 mg or more actin from 6 l of yeast.

4.3.3 Separation of mixtures of wild-type and mutant actins

In some cases, the mutation of interest will not be capable of supporting all of the necessary actin functions in yeast. In these instances, it will be necessary to grow large quantities of diploid cells which contain both wild type and mutant actins. If the mutant protein binds to DNase I, then the two actins can easily be purified. We provide here several suggestions for the next steps in separating the two proteins.

- *Hydroxyapatite chromatography.* Karlsson (40) co-expressed yeast and chicken β-actin, purified the mixture of actins by DNase I affinity chromatography, and then separated the two proteins by hydroxyapatite chromatography. It is possible that different yeast actins could be separated similarly, but this procedure is very sensitive to the particular hydroxapatite preparation used.

- *Sulphydryl affinity chromatography.* Yeast grow normally when expressing a yeast actin in which the Cys at position 373 is changed to an Ala (D. Gallwitz, personal communication). If the mutation of interest is cloned into the coding sequence which contains the Cys → Ala mutation, then the mutant actin will have one less Cys than the wild-type actin. Since this is the only Cys which is accessible, as determined by labelling studies, the wild-type and mutant actins may be separable by the differences in their affinity for an activated thiol column.

- If the mutant of interest does not bind to DNase I, then another technique, such as the Scordilis and Anable protocol (52), must be used to purify the mixture of actins. The two proteins can then be separated by passing the mixture over a DNase I agarose column and collecting the mutant actin in the flow-through.

5. Functional studies

Once the mutant actins have been produced and purified, they can be studied functionally *in vitro* in terms of parameters such as nucleotide binding, metal binding, polymerization, interaction with various actin-binding proteins, and the production of contractile force. These protocols can be found in papers or reviews dealing with these particular aspects of actin function (53–56) and will not be described here. Additional protocols pertaining to actin function in yeast are described in the works by Kilmartin and Adams (57), Adams and Pringle (58), and other works reviewed by Drubin (59).

Acknowledgements

This work was supported in part by a grant from the NIH to P.A.R. (GM-33689) and by an NSF predoctoral fellowship to R.K.C.

References

1. Pollard, T. D. and Cooper, J. A. (1986). *Ann. Rev. Biochem.*, **55**, 987.
2. Levine, B. A., Moir, A. J. G., and Perry, S. V. (1988). *Eur. J. Biochem.*, **172**, 389.
3. Sutoh, K. (1982). *Biochemistry*, **21**, 3654.
4. Doi, Y., Higashida, M., and Kido, S. (1987). *Eur. J. Biochem.*, **164**, 89.
5. Sutoh, K. and Mabuchi, I. (1984). *Biochemistry*, **23**, 6757.
6. Sutoh, K. and Hatano, S. (1986). *Biochemistry*, **25**, 435.
7. Firtel, R. A., Timm, R., Kimmel, A. R., and McKeown, M. (1979). *Proc. Nat. Acad. Sci. USA*, **76**, 6206.
8. Nudel, U. Zakut, R., Shani, M., Neuman, S., Levy, Z., and Yaffe, D. (1983). *Nucleic Acids Res.*, **11**, 1759.
9. Gallwitz, D., and Sures, I. (1990). *Proc. Nat. Acad. Sci. USA*, **77**, 2546.
10. Ng, R. and Abelson, J. (1980). *Proc. Nat. Acad. Sci. USA*, **77**, 3912.
11. Losberger, C. and Ernst, J. F. (1989). *Nucleic Acids Res*, **17**, 9488.
12. Mertins, P. and Gallwitz, D. (1987). *Nucleic Acids Res.*, **15**, 7369.
13. Fidel, S., Doonan, J. H., and Morris, N. R. (1988). *Gene*, **70**, 283.
14. Hamada, H., Petrino, M., and Kakungaga, T. (1982). *Proc. Nat. Acad. Sci. USA*, **79**, 5901.
15. Zakut, R., Shani, M., Givol, D., Neuman, S., Yaffe, D., and Nudel, U. (1982). *Nature*, **298**, 857.
16. Fyrberg, E. A., Bond, B. J., Hershey, N. D., Mixter, K. A., and Davidson, N. (1981). *Cell*, **24**, 107.
17. Kost, T. A., Theodorakis, N., and Hughes, S. H. (1983). *Nucleic Acids Res.*, **11**, 8287.
18. Kabsch, W., Mannherz, H. G., Suck, D., Pai, E. F., and Holmes, K. C. (1990). *Nature*, 347, 37.
19. Holmes, K. C., Popp, D., Gebhard, W., and Kabsch, W. (1990). *Nature*, **347**, 44.
20. Drubin, D. G. (1990). *Cell Motil. Cytoskel.*, **15**, 7.
21. Smith, M. (1985). *Ann. Rev. Gen.*, **19**, 423.
22. Shortle, D. and Nathans, D. (1978). *Proc. Nat. Acad. Sci. USA*, **75**, 2170.
23. Zoller, M. J. and Smith, M. (1982). *Nucleic Acids Res.*, **10**, 6487.
24. Taylor, J. W., Ott, J., and Eckstein, F. (1985). *Nucleic Acids Res.*, **13**, 8764.
25. Kunkel, T. A., Roberts, J. D., and Zakour, R. A. (1987). *Meth. Enzymol.*, **154**, 367.

26. Jones, D. H., and Howard, B. H. (1990). *BioTechniques*, **8**, 178.
27. Hutchison III, C. A., Nordeen, S. K., Vogt, K., and Edgell, M. H. (1986). *Proc. Nat. Acad. Sci. USA*, **83**, 710.
28. Cunningham, B. C., and Wells, J. A. (1989). *Science*, **244**, 1081.
29. Hitchcock-DeGregori, S. E. (1989). *Cell Motil. Cytoskel.*, **14**, 12.
30. Frankel, S., Condeelis, J., and Leinwand, L. (1990). *J. Biol. Chem.*, **265**, 17980.
31. Sussman, D. J., Sellers, J. R., Flicker, P., Lai, E. Y., Cannon, L. E., Szent-Gyorgi, A. G., and Fulton, C. (1984). *J. Biol. Chem.*, **259**, 7349.
32. Johnson, P., Harris, C. I., and Perry, S. V. (1967). *Biochem. J.*, **105**, 361.
33. Collins, J. H. and Elzinga, M. (1975). *J. Biol. Chem.*, **250**, 5915.
34. Asatoor, A. M. and Armstrong, M. D. (1967). *Biochem. Biophys. Res. Comm.*, **26**, 168.
35. Solomon, L. R. and Rubenstein, P. A. (1987). *J. Biol. Chem.*, **262**, 11382.
36. Redman, K. and Rubenstein, P. A. (1981). *J. Biol. Chem.*, **256**, 13226.
37. Rubenstein, P. A. and Martin, D. J. (1983). *J. Biol. Chem.*, **258**, 11354.
38. Ng, R., Domdey, H., Larson, G., Rossi, J. J., and Abelson, J. (1985). *Nature*, **314**, 183.
39. Shortle, D., Novick, P., and Botstein, D. (1984). *Proc. Nat. Acad. Sci. USA*, **81**, 4889.
40. Karlsson, R. (1988). *Gene*, **68**, 249.
41. Konarska, M. M., Padget, R. A., and Sharp, P. A. (1984). *Cell*, **38**, 731.
42. Solomon, T. L., Solomon, L. R., and Rubenstein, P. A. (1988). *J. Biol. Chem.*, **263**, 19662.
43. Fyrberg, E. A. (1989). *Cell Motil. Cytoskel.*, **14**, 118.
44. Karlik, C. C., Coutu, M. D., and Fyrberg, E. A. (1984). *Cell*, **38**, 711.
45. Reedy, M. C., Beall, C., and Fyrberg, E. A. (1989). *Nature*, **339**, 481.
46. Shine, J. and Dalgarno, L. (1974). *Proc. Nat. Acad. Sci. USA*, **71**. 1342.
47. Schleif, R. F. and Wensink, P. C. (1981). *Practical methods in molecular biology*, Springer-Verlag, New York.
48. Ito, H., Fukuda, Y., Murata, K., and Kimura, A. (1983). *J. Bacteriol.*, **153**, 163.
49. Glover, D. M. (1985). *DNA cloning: a practical approach*. Vol. I. IRL, Oxford.
50. Zechel, K. (1980). *Eur. J. Biochem.*, **110**, 343.
51. Greer, C. and Schekman, R. (1982). *Mol. Cell Biol.*, **2**, 1270.
52. Scordilis, S. P. and Anable, D. (1990). *J. Cell Biol.*, **111**, 31a.
53. Pollard, T. D. and Korn, E. D. (1973). *J. Biol. Chem.*, **248**, 4682.
54. Estes, J. E., Selden, L. A., and Gershman, L. C. (1987). *J. Biol. Chem.*, **262**, 4952.
55. Pollard, T. D. (1984). *J. Cell Biol.*, **99**, 769.
56. Kron, S. J. and Spudich, J. A. (1986). *Proc. Nat. Acad. Sci. USA*, **83**, 6272.
57. Kilmartin, J. V. and Adams, A. E. M. (1984). *J. Cell Biol.*, **98**, 922.
58. Adams, A. E. M. and Pringle, J. (1984). *J. Cell Biol.*, **98**, 934.
59. Drubin, D. (1989). *Cell Motil. Cytoskel.*, **14**, 42.
60. Spudich, J. and Watt, S. (1971). *J. Biol. Chem.*, **246**, 4866.
61. Cook, R. K., Sheff, D. R., and Rubenstein, P. A. (1991). *J. Biol. Chem.*, **266**, 16825.

6

Association of cytoskeletal proteins with membranes

CORALIE A. CAROTHERS CARRAWAY

1. Introduction

Interactions of the plasma membrane with the submembrane cytoskeleton (reviewed in refs 1 and 2), particularly with microfilaments (3, 4), have been implicated both in the transduction of information and in its subsequent utilization in cellular processes. Interaction with other cells and with the extracellular matrix, immune recognition, the localization and function of receptors and enzymes, capping of cell surface antigens, the polarity of epithelial cells, endocytosis, and cell division are only a few examples of processes which involve cytoskeleton-mediated specific localizations of cell surface molecules. Examples of some of the types of membrane–cytoskeleton interactions and their postulated roles are given in *Table 1*. Disruption of any of these normal processes of growth, activation, differentiation, and tissue formation can lead to malignant transformation and numerous other disease states. One significant aspect of the problem of the roles of membrane–cytoskeleton interactions in dynamic cellular processes is how microfilaments are linked to plasma membranes. The molecular organization at the membrane–microfilament interface remains poorly understood (1), except in the case of the erythrocyte (5), the simplest and best-characterized cell type. A more complete characterization is developing also for these interactions in intestinal brush border (6), adhesion plaques (7, 8), platelets (9), ascites tumour cell microvilli (in ref. 1), and *Dictyostelium* (10).

A number of ultrastructural and biochemical approaches have been used in the study of membrane–cytoskeleton interactions. Microscopy (Chapters 1 and 2) and physical studies, which are beyond the scope of this chapter, have provided evidence for the localization of proteins in the submembranal or cortical region of the cell, but have not always provided sufficient resolution to establish specific protein–protein interactions. Several biochemical approaches have been taken for these analyses, including analysis of purified plasma membrane preparations for cytoskeletal proteins, co-purification of cytoskeletal and membrane proteins, and reconstitution of complexes from

Table 1. Types of membrane–cytoskeleton interaction and postulated functions

Structure	Membrane	Cytoskeletal element	Putative function
microvilli	plasma	microfilaments	formation, stabilization
patches, caps	plasma	microfilaments, microtubules	receptor mobility, immobilization; signal transduction
adherens junctions	plasma		
macula adherens			stabilization of cell– substratum interactions
		intermediate filaments	
desmosomes			cell–cell junctions
cilia, flagella	plasma	microtubules	motility
axoplasmic vesicles	intraceulluar	microtubules	axoplasmic transport
synaptic vesicles	intracellular	microfilaments	secretion
nuclear membrane	intracellular		
cytoplasmic face		intermediate filaments	
nuclear lamina (inner face)		lamin-type inter- mediate filament meshworks	cell surface–nucleus signal transduction

purified components. As pointed out by Geiger (3), each has its advantages and limitations, and for that reason more than one experimental approach should be used in attempting to establish such associations.

Many of the approaches in the characterization of protein–protein interactions, including associations of proteins with actin (for example, those given in Chapters 3, 4, and 5), are routinely used for the study of membrane protein–cytoskeleton interactions. The focus of this chapter, then, is perspectives, approaches, and biochemical methods currently in use for the identification and characterization of cytoskeleton associations with the plasma membrane, particularly with cell surface proteins. Because the association of microfilaments with plasma membranes is the most thoroughly elucidated, the chapter concentrates on this type of membrane–cytoskeleton interaction, although applications to the study of other types of cytoskeletal proteins with membranes are mentioned. A recent review highlights studies on the association of intermediate filaments with membranes (11). In some cases, representative protocols from specific work are presented; in the cases of commonly used or experimentally simple methods, more general or optimized protocols are given. Review articles, particularly methods reviews, are liberally cited. Because of the special requirements for the manipulation of integral membrane proteins, frequent reference is made to a previous volume of the *Practical Approach* series on membranes (12) and membrane reviews.

Coralie A. Carothers Carraway

1.1 The cytoskeleton

The cytoskeleton is usually said to consist of filamentous cytoplasmic structures of three types. Microfilaments (5–7 nm in width) are made up of actin subunits, of which there are three primary tissue-dependent isoforms (see Chapter 5). Intermediate filaments (10 nm) are both more diverse and more stable than either of the other classes. Intermediate filaments are made up of one or more of six classes of IF subunits, depending on the tissue type (see Chapter 10). Microtubules (25 nm), comprised of α- and β-tubulin, are more labile than the other types of filamentous structures, undergoing rapid and extensive alterations during cell division.

In addition to the filamentous structures, a special class of membrane proteins which interact with the cytoskeleton is also considered to consist of cytoskeletal, or more appropriately, membrane skeletal, proteins.

1.2 Membrane proteins

Membrane proteins are a unique class, having properties which in many cases necessitate special considerations in the design of protocols for their study. The fluid mosaic model for membrane structure propounded in the early 1970s (13) described two classes of membrane proteins which differed with respect to both their disposition at the membrane and the conditions required for their solubilization. Integral membrane proteins intercalate into the phospholipid bilayer and cannot be removed without disrupting the bilayer with membrane-solubilizing agents such as detergents. Almost all of the integral proteins characterized to date are transmembrane proteins, having one or more domains passing through the bilayer; most have a domain in the cytoplasm (reviewed in 14). Such a disposition has important implications for potential transmembrane signalling mechanisms. Peripheral proteins, on the other hand, are associated with integral proteins through protein–protein interactions, or with phospholipid head groups, and can be removed under conditions which alter ionic protein–phospholipid and protein–protein interactions, such as high pH or high salt concentrations.

In recent years a third category of membrane-associated proteins has appeared which complicates the original description. Certain proteins which do not have membrane-spanning domains can associate with the bilayer by one of two apparent mechanisms. One type of protein in this class has a limited hydrophobic segment which apparently interacts with the hydrophobic lipid core. It is thought that these proteins intercalate into the bilayer but do not span it. They are less readily removed from the membrane by agents which disrupt ionic interactions. Many of these proteins bind in *in vitro* studies to the hydrophobic portion of phospholipid vesicles, showing no head-group specificity. Another type of protein in this non-integral, non-peripheral class interacts with the bilayer through covalent lipid post-translational modifications. These behave operationally as integral proteins, i.e. they cannot be extracted under conditions used for peripheral protein extraction; they can be removed only by detergent extraction or by cleavage of a bond between the lipid moiety and the protein. An increasing number of proteins are being shown to have covalently attached lipid (reviewed in 15–17). Some of these lipid-modified proteins are at the cell surface (17), but many are associated with the cytoplasmic face of the membrane, or have been localized to the cortical (sub-plasmalemma!) region of the cell. Some of these proteins exist in either soluble or membrane-associated forms, and can be recruited to the plasma membrane after a signalling event. The potential regulatory importance of such signal-mediated translocations makes the characterization of the lipid modifications and the study of interactions of these membrane proteins with membrane lipid and with proteins of both the membrane and the cytoskeleton a likely target area for future studies.

A number of the proteins in this third class, termed amphitropic proteins by Burn (18), are considered to be cytoskeletal proteins, especially if they are known to interact either directly or indirectly with filamentous cytoskeletal structures. Assemblages of membrane and cytoskeleton proteins have been called the membrane skeleton or membrane scaffold. A schematic represent-ation of the three classes of membrane proteins is shown in *Figure 1*.

2. Approaches to the biochemical characterization of membrane–cytoskeleton interactions

Two basic approaches taken for the biochemical characterization of association of membrane proteins with cytoskeletal components are co-isolation and reconstitution. Co-isolation is a reasonably reliable indication of a physio-logically significant association but is frequently not feasible because either

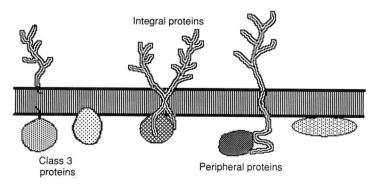

Figure 1. A schematic representation of the three classes of membrane proteins

the interactions are transient, as after a signalling event, or the avidities of the components for one another are insufficient to withstand a lengthy purification procedure. When an interaction is sufficiently stable, multiple components may be present in the complex, preventing identification of the components which interact directly. Reconstitution of isolated components gives some indication of the likelihood (or lack of) of an *in vivo* interaction, but is insufficient in itself to substantiate a possible physiological role. This is especially true if the *in vitro* interaction occurs preferentially under non-physiological conditions. Another problem with reconstitution studies is that not all of the constituents which modulate a protein–protein interaction may be present in the *in vitro* assay. A third difficulty is that some interactions thought to be physiologically important are apparently weak associations, necessitating an experimental design which takes this problem into account. Ideally, biochemical demonstration of specific interactions includes both co-purification and reconstitution studies, where feasible.

2.1 Initial identification of membrane proteins associated with cytoskeletal components

Fundamentally different approaches are taken to identifying unknown cell surface proteins, largely glycoproteins, and peripheral cytoplasmic proteins associated with the cytoskeleton. Integral membrane proteins can be identified by methods which are specific for cell surface proteins or for glycosylated proteins. In the case of uncharacterized microfilament-associated integral membrane proteins, a number of methods have been used *in situ* for either chemical modification and isotopic labelling of the cell surface exclusively or specific labelling of the carbohydrate moieties of the glyco-proteins. Staining of gels after SDS-PAGE with carbohydrate-specific stains has also been used. Some of these methods for glycoprotein detection are summarized in *Table 2*. Each method has its advantages and inherent drawbacks. An extensive review of many of these methods is given in ref. 19, and protocols for some have been presented in ref. 12 (Chapter 6).

Cytoplasmic peripheral proteins, which contain no carbohydrate and cannot be labelled in this manner, require techniques more specific for the individual protein, such as immunological identification or biochemical characterization of the purified protein. General methods for integral proteins will be emphasized in this chapter, since methods for non-glycosylated proteins include many of the common approaches for the study of protein–protein interactions. Some of the most useful of these general techniques for the study of actin-binding proteins are described in Chapter 4 of this volume. For peripheral proteins, representative examples of routine approaches for studying protein–protein interactions which have been used for characterizing the interaction of a specific membrane protein with a cytoskeletal protein will be given.

Table 2. Methods for labelling membrane glycoproteins

Method	Advantages	Disadvantages	Refs.
periodate/Schiff staining of gels	stains glycosylated proteins only	insensitive; stains heavily sialated proteins selectively; fades rapidly	21
metabolic labelling with glucosamine, mannose, or fucose	mild labelling method, will not alter structure or function; labels only endogenous proteins	may not be sensitive enough for low-abundance proteins	20
solid-phase radio-iodination of cell-surface proteins	sensitive; relatively non-selective	somewhat harsh conditions; modification may alter binding properties; high background; handling, disposal problems; high cost	2, 19
galactose oxidase-[^{3}H]borohydride,	sensitive, relatively non-selective	not sensitive enough for non-abundant proteins; can potentially cross-link proteins	19, 22
lectin binding	can be made very sensitive; no isotope required; can be performed on gels or blots; can be used to identify specific classes of glycoproteins	may not detect poorly glycosylated proteins	23

2.1.1 Co-sedimentation after membrane solubilization

The first step in identifying integral membrane proteins associated with the cytoskeleton is solubilization of the plasma membrane of intact cells with non-ionic detergent, digitonin, or saponin to release varying amounts of the membrane lipid and all soluble components. Descriptions of considerations in the selection and use of detergents for membrane solubilization are given in refs 12 (Chapter 5), and 24 (Chapter 1). Considerations in the choice of detergent include critical micelle concentration (detergents having high critical micelle concentrations for dialysability), mildness (for retention of enzymatic activities or binding functions), and price (the cost of some precluding their use in large volumes, as in chromatographic separations).

Other important factors to consider in detergent extraction include:

- Concentration of detergent (or, more accurately, detergent:lipid ratio, more frequently determined for a particular membrane or cell system as detergent:protein ratio)
- Ionic composition of the buffer
- Incubation time
- Incubation temperature.

To retain weak associations, membranes should be solubilized in the mildest detergent in a physiological buffer for the briefest possible time. A preliminary extraction study examining these parameters is a requisite first step in any biochemical characterization of membrane–cytoskeleton interactions.

After solubilization, a simple pelleting procedure is normally performed, resulting in a 'cytoskeletal residue' which may then be assayed for the presence of glycoproteins. Considerations in the pelleting of actin filaments are given in Chapter 3 of this volume. If the cell system in question is amenable to metabolic labelling with radiolabelled glucosamine, the pellet can be simply assayed by SDS-PAGE and fluorography. Mannose labelled at the 2-position and fucose can also be used for labelling, but glucosamine is preferable for survey experiments because it is incorporated into glycoproteins of all classes. General considerations in metabolic labelling of glycoconjugates may be found in ref. 20.

2.1.2 Phalloidin shift on velocity sedimentation gradients for identification of microfilament-associated proteins

A difficulty with the pelleting assay for association of cell surface proteins with the cytoskeleton is that the insoluble residue from cell extracts is a complicated mixture of organelle remnants and all types of cytoskeletal structures. After a pelleting experiment has suggested the presence of cytoskeleton-associated glycocoproteins, a more careful analysis is needed.

The next step is to analyse either intact cells, or, preferably, a stabilized plasma membrane preparation which retains associations with the cytoskeleton. We have reviewed membrane preparation methods, including stabilized membranes, for a variety of cell types (25); some basic protocols for cell disruption and membrane fractionation have been described (12, Chapter 1). Isolated microvilli from ascites tumour cells, prepared under conditions which do not disrupt the cell body (26), are a highly purified, minimally perturbed plasma membrane preparation retaining its associated microfilaments (27). Microvilli have also been prepared by this basic procedure from chondrocytes (28). Many other cells also have microvilli, and a modification of this procedure, perhaps including a stabilization step, such as inclusion of Zn^{++} or borate in the shearing buffer, might be used in microvillus isolation from some of these.

The stabilized membrane preparation is then extracted, fractionated and analysed for complexes of membrane and cytoskeleton proteins. Gel filtration chromatography or velocity sedimentation centrifugation rather than a simple pelleting procedure may be used for fractionation, since these methods allow the possibility of resolution and identification of specific complexes (for an example, see ref. 29). When used in conjunction with a stabilizer of a specific cytoskeletal structure, velocity sedimentation can provide a diagnostic first analysis of the association of cell surface proteins with that filament type (30). Moreover, filament fractions from a gradient may serve as a starting point for purification of the complex or a complex component.

An example of this approach is the phalloidin shift on velocity sedimentation technique for the identification of microfilament-associated proteins (30). Velocity sedimentation fractionations are performed on extracts of cells, or, preferably, a stabilized plasma membrane fraction, in the presence and absence of phalloidin, a mushroom toxin which specifically stabilizes microfilaments. Specific microfilament-associated membrane proteins are very simply demonstrated by this technique when appropriate methods for their detection are available. Immunoblots, if antibody to a known membrane protein is available, or enzymatic (31) or receptor ligand-binding assays are ideal diagnostic methods for specific proteins of interest. Metabolic labelling with glucosamine, if feasible in the cell of interest, is advantageous, since it is a non-perturbing diagnostic method for cell surface glycoproteins (29, 32). *Protocol 1* describes the basic method for phalloidin shift velocity sedimentation analysis of the isolated microvilli metabolically labelled with [^{14}C]glucosamine.

Protocol 1. Phalloidin shift on velocity sedimentation gradients for analysis of microfilament-associated cell surface proteins

1. Label cells metabolically by adding 5 μCi [^{14}C]glucosamine (ICN) per 10^6 cells in sterile solution. Incubate for 4–16 h to incorporate label (some proteins turn over more slowly than others).

2. Solubilize two aliquots of cells or membrane fraction in a total volume of 1.0 ml of 0.2 per cent Triton X-100, 150 mM KCl, 2 mM MgCl$_2$, and 20 mM Pipes, pH 6.8 (TPK buffer) plus the protease inhibitor cocktail given in Chapter 5, without and with 10 μg phalloidin for 15 min at 4 °C.

   • Similar results are obtained in buffer containing octylglucoside, which is more easily dialysable but much more expensive than Triton.

3. Immediately layer extracts on to linear 20–40 per cent sucrose gradients in the Triton-containing buffer in SW40 or SW40.1 tubes (13–14 ml).

4. Centrifuge the gradients at 4 °C and 30 000 rev/min for 3 h in an SW40 or SW40.1 rotor. Collect 0.5 ml fractions by extrusion from the tops of the tubes into calibrated microcentrifuge tubes.

5. Determine sucrose densities by refractive indices on 20 μl aliquots of vortexed gradient fractions.

6. Obtain carbohydrate profiles by scintillation counting of 200 μl aliquots of the fractions.

7. To the remainder of each gradient fraction (about 750 μl) add SDS to 1 per cent and prepare for electrophoresis by dialysis against 0.1 per cent

SDS/0.1 mM EGTA/0.02 per cent NaN$_3$ (8–9 h) at room temperature, three changes of dialysis medium.

- Addition of SDS to samples prior to dialysis accomplishes two objectives, displacement of Triton bound to the proteins and facilitation of solubilization after lyophilization. Many proteins, particularly membrane proteins and complexes, do not resolubilize readily after lyophilization in the absence of denaturant.

8. Freeze the dialysed samples immediately in a dry ice–acetone bath and lyophilize. Solubilize in electrophoresis buffer immediately prior to electrophoretic analysis on 8 per cent or 6–12 per cent gradient SDS-PAGE gels (not mini-gels, if fluorography is to be done).

- In our experience, gels which have been polymerized for $\geq$ 12 h give much better resolution and sharper glycoprotein bands.

9. After staining the gels with Coomassie Blue, prepare them for fluorography. Exposure time is determined by the abundance of the glycoproteins and the extent of label incorporation.

- Minimize the time of gel exposure to organic solvents, as many glycoproteins appear to be extractable in alcohols.

It cannot be over-stressed that the composition of the extraction buffer and the extraction conditions are very important to the success of any biochemical approach to the study of protein–protein interactions. Inclusion of a protease inhibitor cocktail in buffers during fractionations is critical, as discussed in Chapters 5 and 10, especially when disrupting and fractionating whole cells (12, 24). The ionic composition of the buffer is also very important to the stability of protein–protein interactions. The TPK buffer used for extraction of our tumour cell microvilli was selected after a careful analysis of optimal conditions for maintenance of the integrity of the ascites tumour cell microvillar microfilament meshworks. We have recently found that replacement of Triton with octyl glucoside in the same buffer retards fragmentation even better.

The phalloidin shift protocol might be readily modified for the analysis of microtubule-associated proteins by use of an optimal microtubule-stabilizing buffer (33) for extraction with and without a specific microtubule-stabilizing agent such as taxol.

The highly cross-linked microfilament meshworks in our tumour cell microvilli make them easy to sediment. Depending on the size and stability of the microfilament meshworks and bundles in a particular cell type, and especially for non-filamentous actin found in stabilized membrane preparations, the centrifugation conditions may need to be modified for optimal sedimentation. For example, in analysing neuroblastoma cells by this method, we used a 7–30 per cent sucrose gradient and increased the centrifugation time to 16 h (C. Carraway and Y. Palacio, unpublished observations). Although it was not necessary to include phalloidin in the gradient buffers for the short centrifugation times, it was included for the longer times.

Glycoprotein distribution patterns (from either metabolic or exogenous isotopic labelling), when compared to gels stained with Coomassie Blue, give an indication of the presence and amounts of glycoprotein in the filament fractions. Label found in the phalloidin-shifted as well as the control microfilament fractions can be more confidently interpreted as being associated specifically with those filaments. Fluorograms of the SDS-PAGE gels can then be performed to identify specific microfilament-associated glycosylated proteins for further study (32).

A more convenient and potentially powerful method of analysis for cell surface proteins which obviates the necessity for radiolabel is the lectin blot, a form of Western blot which uses lectin rather than antibody as the ligand, and a variation of the lectin overlay on gels (23). Properties of lectins, their carbohydrate specificities, and their uses in the study of glycoproteins have been extensively reviewed (34–36). The basic methods for electrophoresis on minigels and transfer of proteins to blots, described briefly by Matsudaira in Chapter 4 of this volume, are discussed in detail in a number of methods volumes, including ref 24 (Chapter 8) and 37 (Chapter 20). Nitrocellulose is the blot matrix of choice, although nylon and polyvinylidene difluoride (PVDF) are also used for this purpose. Their relative advantages and disadvantages are discussed in ref. 24, and methods specifically for PVDF membrane blotting are given in ref. 38. Nitrocellulose is less expensive than nylon membrane, and the chemical resistance offered by PVDF membranes is not required unless transferred proteins are to be sequenced. PVDF membranes, like nylon, give routinely higher background staining. In our experience nitrocellulose membrane by Schleicher and Schuell is better for the transfer of high M_r proteins, but Gelman Biotrace NT binds intermediate M_r proteins, particularly glycoproteins, better. The Gelman product, which also gives a cleaner background, is less brittle and thus easier to handle.

The choice of lectin for the blot can be based upon knowledge of the types of glycosylation on specific presumed cytoskeleton-associated glycoproteins or, lacking information relating to types of carbohydrate structures on the proteins of interest, a brief survey using lectins recognizing common carbohydrate moieties in glycoproteins. *Table 3* lists lectins having specificities for carbohydrate moieties found commonly on membrane glycoproteins. Lectin blots have been used to identify glycoproteins associated in transmembrane complexes with cytoplasmic components. Concanavalin A (Con A) blots and fluorography of metabolically labelled proteins were used to show the presence in isolated tumour cell microvillar membranes of a very large transmembrane complex containing at least four glycoproteins (32) in a

Table 3. Lectins having specificities for carbohydrate structures found on membrane glycoproteins

Lectin	Sugar specificity	Comments
D-mannose-binding		
concanavalin A (Con A)	branched α-linked mannoses and biantennary N-linked oligosaccharides	recognizes high mannose type glycoproteins frequently found in plasma membranes
N-acetyl-glucosamine-binding		
wheat germ (WGA)	$(D\text{-glcNAc})_{2 \text{ or } 3}$ (including terminal); >> NeuNAc	recognizes structures commonly found on both *N*- and *O*-glycosylated serum and membrane proteins
Complex oligosaccharide-binding		
Triticum vulgaris (PHA-L)	multi-antennary complex type oligosaccharides	recognizes a relatively rare structure which is correlated with differentiation changes
tomato lectin (LEA)	polylactosamine	specific for polylactosamine moieties, which have been implicated in malignant transformation and metastasis
Galactose-binding		
ricin (RCA I and II)	ß-D-gal-containing complex oligosaccharides without sialic acid	recognizes both *N*- and *O*-linked oligosaccharides; very toxic
peanut agglutinin (PNA)	ß-D-gal-ß(1,3)galNAc (T-antigen); does not bind this moiety when terminal sialic acid is present	recognizes sugar moieties specific for *O*-linked sugars (used + and − sialidase treatment)
jacalin	T-antigen with terminal sialic acid	not as well characterized as PNA, recognizes a larger variety of *O*-linked oligosaccharides

stable association with actin and a 58 kDa protein. Wheat germ agglutinin (WGA) blots, shown in *Figure 2*, were used to show the association of glycoproteins in skeletal muscle cell membranes with dystrophin (39), a cytoplasmic peripheral membrane protein thought to be linked to the cytoskeleton. The basic method for performing lectin blots is given in *Protocol 2*. Lectin blots may also be performed after transfer from two-dimensional isoelectric focusing-SDS-PAGE gels (4).

Protocol 2. General method for lectin detection of glycosylated membrane proteins

1. Run fractions containing membrane-associated cytoskeletal components on SDS-PAGE gel of an appropriate acrylamide concentration.

2. Equilibrate the gel for 15 min in transfer buffer (20 per cent methanol, 0.025 M Tris, 0.025 mM glycine, pH 9.5); transfer the proteins from the gel to nitrocellulose paper, using 20 V overnight (room temperature) or 100 V for 1.5 h in the cold room. It is important to establish optimal transfer conditions for the proteins of interest.

3. Block non-specific sites with 3 per cent BSA/20 per cent Tween-20, 0.5 M NaCl, 0.05 M Tris, pH 8.0 (TTBS) at room temperature for 1 h.

4. Incubate with lectin–horseradish peroxidase (HRP) or lectin–alkaline phosphatase (AP) (Sigma), diluted 100-fold with 0.3 per cent BSA/TTBS for 1 h at room temperature with shaking.

   ● Save this solution; it can be used up to three times.

5. Wash the blot with TTBS three times, 10 min each.

6. Rinse with 0.5 M NaCl, 0.05 M Tris, pH 8.0, and develop with the appropriate colour reagent, prepared just before use, as described in the literature accompanying the colour development reagents.

The choice of AP-conjugated or HRP-conjugated lectins should be determined by several offsetting considerations:

AP-conjugation of ligands provides:

● greater sensitivity than HRP-conjugated ligands in the conventional colour development procedures

● better colour development product stability than the HRP-conjugated ligands

HRP-conjugation affords:

● considerably less expense than AP-conjugated ligands

● faster enzymatic turnover rates, yielding a faster accumulation of product.

A very convenient and more sensitive alternative to the conventional HRP- and AP-colour development systems is provided by two new spray reagents, one for HRP- and one for AP-conjugated lectins, which have recently been marketed (TSI). Opti-Mist HRP is three times as sensitive as the conventional HRP colour development method, and Opti-Mist AP is at least equivalent in sensitivity to the conventional AP method. We have found it necessary to dilute the HRP-conjugated ligands 5–10-fold and the AP-conjugated ligands two-fold prior to incubation of the blots. Another advantage of the Opti-Mist colour development method is that the colours do not fade, as frequently seen with the conventional methods. Both reagents have a long shelf life (12–15 months either refrigerated or at room temperature). One potential problem with this method is that if the concentration of the glycosylated proteins is too high, a thick deposit of stain develops on the blots and will flake off. For this reason a sample should be loaded in more than one concentration, using 0.1–0.5 as much as would be used for the conventional method.

These methods have the sensitivity to detect glycoproteins of low abundance in membranes, and modifications of the technique (41–43) allow for their quantification. Lectins labelled with ^{125}I may be used for both sensitive detection and quantification (see ref. 44 for iodination of Con A). A

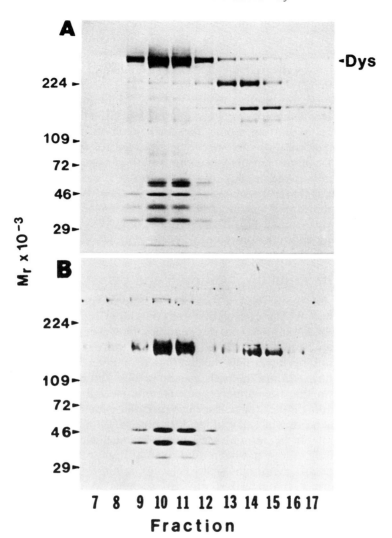

Figure 2. Gels stained with Coomassie Blue (*A*) and with WGA (*B*) of velocity sedimentation sucrose density gradient fractions containing the dystrophin–glycoprotein complex (remounted from photos of panels a and b for Fig. 1, ref. 36, with permission)

useful method for radio-iodination of proteins is described in Chapter 4 of this volume; more extensive descriptions of methods for iodination may be found in refs 12 (Chapter 6) and 37 (Chapter 36). Other methods for labelling lectins, e.g. fluorescence labelling, may be found in ref. 35.

It should not be surprising that intensities of a given glycoprotein band differ between patterns obtained by fluorography of glycoproteins metabolically labelled with glucosamine and by blotting with lectins. For instance, Con A binds to N-glycosylated proteins of high mannose type, but not of complex type. In contrast, glucosamine is more abundantly incorporated into glycoproteins having more complex oligosaccharides. A simple comparison of glucosamine and lectin blotting results can give preliminary information on the types of glycosylation of the cytoskeleton-associated proteins (32), which might serve as a starting point for further characterization of these glycoproteins. Selection of very specific lectins (see ref. 35 for specificities), combined with a more sensitive detection method (1–10 ng glycoprotein) using digoxigenin-linked lectins and antibody to digoxigenin (45) allows rather extensive characterization of carbohydrate structures on proteins without the laborious procedures which have been conventionally used. Given the growing evidence for the importance of glycosylation to membrane protein targeting and function (46–49), glycosylation studies will be an area of increasing focus in the study of membrane protein structure and function.

2.2 Co-purification of membrane and cytoskeleton proteins

One criterion for the *in vivo* association of a cytoskeletal protein with a membrane protein is their co-isolation in a complex. Unless the interaction is particularly stable, this may not be feasible. However, in practice many interactions have been demonstrated by simple affinity techniques which are not lengthy, thereby increasing the likelihood of detection of transient or weak interactions. Affinity methods usually involve solubilization of the cell or membrane in question under gentle conditions and co-isolation of the complex by binding either the cell surface protein or the cytoskeletal protein to a specific ligand to which the other does not itself bind. Affinity methods include both soluble ligand and solid phase-bound ligand techniques (50). This kind of methodology has been used successfully for the co-isolation of a sizable number of membrane–cytoskeleton complexes. When studying interactions with a characterized protein, such as a specific cytoskeletal protein or receptor, an obvious approach to retrieving the complexes is via antibody to one of the components, if monospecific antibody is available. A representative immunoprecipitation procedure is given in *Protocol 3*, a demonstration of the association of two platelet cell surface glycoproteins (gpIb and gpIa) with actin-binding protein (51). In an extension of this procedure, immunoprecipitation with a monoclonal antibody to actin-binding protein has allowed identification of a large fragment of actin-binding protein containing the binding domains for both actin and gpIb (52).

Protocol 3. Immunoprecipitation with anti-gpIb for co-isolation of gpIb, gpIa, and actin-binding protein in platelets

1. Sediment human platelet-rich plasma at 730 × *g* for 10 min; wash the platelet pellet twice in buffer A (120 mM NaCl, 13 mM trisodium citrate, 30 mM dextrose, pH 7.0).

2. Wash once in buffer B (150 mM NaCl, 10 mM Hepes, 1 mM EDTA, pH 7.6). For labelling by sodium metaperiodate/[^{3}H]NaBH$_4$, suspend platelets at 1×10^9 cells/ml in buffer B containing 10 ng/ml prostacyclin.

   - General protocols for labelling cells or membranes with sodium [^{3}H]borohydride after galactose oxidase treatment are given in refs 12 (Chapter 6) and 19.

3. Incubate labelled platelets ($1–2 \times 10^9$) at 37 °C for least 1 h in Tyrode's buffer (buffer C, 138 mM NaCl, 2.9 mM KCl, 12 mM NaHCO$_3$, 0.3 mM sodium phosphate, 5.5 mM glucose, 1.8 mM CaCl$_2$, 0.4 mM MgCl$_2$, pH 7.4) containing 10 ng/ml prostacyclin.

4. Lyse platelets by addition of an equal volume of ice-cold buffer D (2 per cent Triton X-100, 10 mM EGTA, 2 mg/ml leupeptin, 100 mM Tris–HCl, pH 7.4), containing 2 mg/ml DNAse I treated with 1 mM phenylmethane-sulphonyl fluoride (PMSF)).

5. Remove any residual actin filaments by centrifugation of the lysate at 100 000 × g for 3 h at 4 °C.

6. Wash an aliquot of *S. aureus* Protein A-agarose (IgSorb, The Enzyme Center, Inc.) five times in a 1:1 (v/v) mixture of buffer C and buffer D, centrifuging at 15 000 × g for 1 min. Remove proteins binding non-specifically to protein A by incubating the Triton-soluble platelet fraction at 4 °C for 30 min with 0.1 volume of a 10 per cent solution of washed protein A-agarose and centrifuging.

7. Add actin binding protein antibody to a final concentration of 20 µg/ml or control rabbit IgG to the platelet supernate and incubate for 1 h at 4 °C.

8. Centrifuge the mixtures at 15 000 × g for 1 min. Solubilize the pellet with SDS electrophoresis sample buffer and analyse by SDS-PAGE and fluorography.

Antibody to one of the proteins of interest can also be attached to a solid support such as Agarose and extracts loaded onto the resultant immunoaffinity column for retrieving complexes, as well as for purifying the protein. Detailed protocols for immunoaffinity methods are given in ref. 37.

If the membrane–cytoskeleton interaction is transient or weak, chemical cross-linking prior to extraction can be used to stabilize the interaction and allow co-purification. This method is not always successful, since many cross-linkers investigated tend to give very large aggregates of neighbouring proteins (reviewed in reference 19). However, cross-linking can be useful in less complex systems, as evidenced by its successful use in demonstrating an interaction between platelet gpIIb/IIIa with actin (53).

When the nature of the membrane–cytoskeleton association is uncharacterized, or when antibody is not available, more general ligands are used. Widely applicable approaches which have been used to co-isolate complexes of cytoskeletal proteins with cell surface proteins include lectin binding for specific precipitation of cell surface proteins and specific actin-binding methods. The advantage of the co-precipitation approach, when it is feasible, is its simplicity and speed. It should be remembered that buffer composition is very important to the stability of many interactions, and the assay buffer may need to be altered appropriately (e.g. made isoionic or have metal ions added), if these alterations are compatible with the assay.

Protocol 4 illustrates the use of myosin, which binds actin specifically, to co-precipitate H-2 antigen of the P815 murine mastocytoma line with actin (54).

Protocol 4. Myosin precipitation of a membrane–cytoskeletal protein complex

1. Dilute purified rabbit skeletal muscle myosin, stored at $-20\,°C$ in 0.6 M NaCl and 50 per cent glycerol, into 10 mM Tris-HCl, pH 7.5, to a final concentration of 1 mg/ml and incubate for 15 min at 4 °C.

2. Pellet the filaments at $1000 \times g$ for 5 min; wash in Tris buffer and resuspend to 1.0 mg/ml in 10 mM Tris-HCl, pH 7.5, with 1 per cent Nonidet P-40 (NP-40).

3. Pellet 50 µg of myosin filaments and mix with membrane test samples in 100 µl of Tris/NP-40 buffer; incubate for 1 h at 4 °C.

4. Pellet the filaments at $1000 \times g$ for 4 min; assay by SDS-PAGE and lectin blot, or other appropriate method for detection of the cell surface proteins.

If a specific ligand or monospecific antibodies to one of the components are not available, more general approaches, such as lectin affinity chromatography (35), may be used for the isolation of membrane–cytoskeleton complexes. When seeking complexes by this method, appropriate considerations must be kept in mind. Care must be taken in the use of metal chelators, since many multisubunit lectins require divalent cations for stability (35). An important *caveat* is that a given lectin is likely to bind many glycoproteins, and a subsequent step or independent approach is necessary to separate those which are associated with the cytoskeletal protein from the non-complex glycoproteins. An example is the demonstration by wheat germ agglutinin–Sepharose of an association of dystrophin with cell surface glycoproteins (55) and the subsequent isolation of a glycoprotein–dystrophin complex by velocity sedimentation centrifugation (39; shown in *Figure 2*).

Another conventional method for isolation of complexes is gel filtration chromatography. Since velocity sedimentation separations are based on hydrodynamic properties, gel filtration offers an alternative separation approach based on size. Both velocity sedimentation and gel filtration have been used for the isolation of a transmembrane complex containing actin stably associated with a cell surface glycoprotein from tumour cell microvilli (29). However, conventional gel filtration, like velocity sedimentation, is slow and thus not useful in studying weak but possibly physiologically significant interactions. A modification of the procedure which allows the study of weak associations is described in *Protocol 7*.

2.3 Reconstitution of membrane and cytoskeleton protein complexes

A variety of approaches have been taken for studying associations of membrane and cytoskeletal proteins (cf. 56). One common approach to studying the interaction of two proteins is affinity chromatography (50), discussed in Section 2.2 as a method for isolation of complexes. This method as applied to reconstitution entails purification and attachment of one of the proteins of interest to a solid support. An extension of this approach is being used to identify proteins binding to specific domains by the attachment of protein fragments to a solid support. As examples, the affinity chromatography method using synthetic peptide corresponding to the cytoplasmic domain of β-integrin (CSAT) was used to demonstrate linkages with α-actinin (57) and with fibulin (58).

2.3.1 Reconstitution of purified proteins

The ultimate test for interaction of proteins is reconstitution of the purified components. *In vitro* binding assays have been important in the elucidation of many types of interactions, e.g. hormone–receptor interactions and the association of cytoskeletal components. Ideally, if the interaction is strong enough, one should be able to observe binding between two proteins under physiological conditions, taking into consideration also any other required components which might be present transiently.

The essential element of a binding technique is that it allow separation of bound and free ligand, usually done by methods which take advantage of differences in either size or sedimentation rate of the bound and unbound ligand. A second requirement is a minimal level of non-specific binding to any component of the system. Many such procedures are available for binding studies, including equilibrium dialysis, velocity sedimentation, gel filtration, and rapid filtration on differentially permeable matrices. Equilibrium dialysis, invaluable in the study of binding of small ligands to proteins, cannot be used for a ligand as large as a protein. Numerous methods optimized for the properties of the proteins of interest have been developed.

Some of the most frequently used techniques in the study of membrane–cytoskeleton interactions are those employed in the studies of actin with proteins. A number of these are described in Chapters 3 and 4 and in other methods volumes and reviews (59–61) and will not be belaboured here, except for the examples given below. A sampling of methods for labelling and using actin in binding studies are listed in *Table 4*. Some of these methods are clearly applicable to the interaction of proteins with cytoskeletal proteins other than actin. For example, the G-actin overlay method described in Chapter 4 has been modified to assay tubulin-binding proteins (62). F-actin overlays (77) and F-actin beads (76) have been used to identify an actin-binding, developmentally regulated glycoprotein in *Dictyostelium*.

Table 4. Labelling of actin and actin-binding methods

	Method	References
Labelling of actin	^{125}I	63
	fluorescence	64–66
	biotin	67
Actin binding		
G-actin	gel overlay	68, 69
F-actin	stabilized F-actin, preparation	70–72
	F-actin columns	73–75
	F-actin beads	76
	F-actin overlay	77

Direct binding of a protein to a filamentous cytoskeletal element such as microfilaments may be easily performed by co-sedimentation. This simple approach can be extended to examine the modulation of interactions of cytoskeletal proteins by binding of a membrane protein, as described in *Protocol 5* for studying modulation by protein 4.1 of the spectrin–actin interaction (79).

Protocol 5. Co-pelleting with F-actin to study spectrin–actin–4.1 interaction

1. Solubilize purified spectrin, protein 4.1, and G-actin separately in 150 mM KCl, 2 mM $MgCl_2$, 0.2 mM DTT, 0.2 mM PMSF, 20 TIU aprotinin/1, 10 mM imidazole, pH 7.5.

2. Add G-actin prepared from muscle (59, 80) to a solution of spectrin, one of protein 4.1, and one containing both, in a total of 200–250 µl. Final protein concentrations are 250 µg/ml actin, 200 µg/ml spectrin, and 90 µg/ml protein 4.1. Maintain the same order of addition and comparable mixing for all assays.

3. Incubate for 45 min at room temperature, gently shearing several times during the incubation period by pipetting back and forth.

4. Centrifuge at 100 000 × *g* for 1 h at 15 °C to sediment filaments.

5. Remove the supernatants and resupend the pellets in the same volume of SDS electrophoresis sample buffer.

6. Perform SDS-PAGE of the supernatants and pellets on 5–15 per cent polyacrylamide gels. Scan the stained gels using a densitometer (Model 150 Scanning Densitomer, BioRad Labs) and quantify the areas under the peaks using an electronic planimeter (model 1224 Electronic Digitizer, Numonics Corp.).

  • Accurate quantitation may also be obtained by cutting and weighing the peaks from the densitometer trace.

Velocity sedimentation on sucrose density gradients is an excellent method for the separation of bound and unbound species if the avidities of the proteins are great enough and the interactions are stable under varying sucrose concentrations. An example of the use of velocity sedimentation to study the association of vinculin and talin (81), found in focal contacts, is given in *Protocol 6*.

Protocol 6. Velocity sedimentation for the demonstration of an interaction between vinculin and talin

1. Incubate separately the purified proteins and a mixture of the two, as well as a solution of standard proteins of the appropriate sizes, for 7 h on ice in a total of 0.5–1.0 ml of 150 mM NaCl, 0.1 per cent β-mercaptoethanol, 0.1 per cent sodium azide, 50 mM Tris-HCl, pH 7.5.

2. Layer each solution over a cold 5–20 per cent linear sucrose gradient prepared in the same buffer in Beckman SW40 or SW41 tubes. Centrifuge at 4 °C for 20 h in a Beckman SW40 or SW41 rotor.

3. Collect 0.5 ml fractions; dialyse and analyse by SDS-PAGE.

An association between the two proteins was indicated by a shift in the migration of the proteins in the vinculin–talin mixture to lower fractions in the gradient than either of the separate proteins.

Another frequently used method for studying protein–protein interactions is gel filtration. A modification of the gel filtration method which allows the study of weak interactions (possibly regulated by other cellular components), is equilibrium gel filtration, which is basically a solid-phase modification of the equilibrium dialysis method. An advantage of equilibrium gel filtration is that quantification of the interactions can be performed (82). A disadvantage is that a large amount of the ligand protein is required. *Protocol 7* illustrates the use of equilibrium gel filtration to study the weak transmembrane interaction between talin and the fibronectin receptor CSAT (83; experimental details in 84). The equilibrium gel filtration method successfully identified an association between CSAT and talin, whereas previous experiments using CSAT or talin overlays on blots and velocity sedimentation sucrose density gradient centrifugation had not detected the interaction.

Protocol 7. Equilibrium gel filtration for studying the binding of CSAT and talin

1. Prepare an Ultrogel AcA22 column (0.2×30 cm), having a void volume of 0.54 ml and an included volume of 1.5 ml, in buffer A (20 mM NaCl, 0.1 mM EDTA, 0.1 mM EGTA, 0.1 per cent β-mercaptoethanol, 0.1 per cent NP-40, 20 mM Tris-HCl, pH 7).

2. Load a void volume equivalent (0.54 ml) of CSAT, metabolically labelled with [^{35}S]methionine, on to the column. Elute the antigen at a flow rate of 0.7 ml/h with buffer A, collecting 60 µl fractions.

3. Count 50 µl aliquots for radioisotope. Reserve the remainder for SDS-PAGE.

4. Incubate a mixture of CSAT (400 µg/ml) and talin (200 µg/ml) in buffer A for 30 min. Load a void volume equivalent of the mixture on to the column; elute with buffer, collect 60 µl fractions and count a portion, as for CSAT alone. The void volume fraction (0.54 ml) corresponds to the antigen–ligand complex, and the later fractions to free antigen.

5. Run the remainder of the fractions on 7 per cent non-reducing SDS-PAGE gels (to maintain the disulphide-linked CSAT dimer) and silver stain.

In adapting this procedure for studying weak interactions between other proteins, keep in mind optimizing the buffer composition and the gel percentage for the proteins of interest.

2.3.2 Binding of purified proteins to natural membranes

Reconstitution of complexes of purified membrane and cytoskeletal proteins is not difficult in cases of avid binding. In some cases it is clear that the presence of an intact bilayer enhances the binding capability. A potentially useful model ,system for examining phospholipid-mediated binding of cytoskeletal proteins to integral membrane proteins is a membrane 'stripped' of its peripheral proteins by treatment at high pH. Luna and co-workers have demonstrated an association between exogenous F-actin and integral membrane glycoproteins of *Dictyostelium* membranes prepared in this manner (85). An optimized procedure for the preparation of 'stripped' membranes and demonstration of the association of actin with integral membrane glycoproteins is given in *Protocol 8*.

Protocol 8. Association of F-actin with glycoproteins of membranes stripped of peripheral membrane proteins by high pH treatment

Preparation of 'stripped' membranes

1. Suspend cells ($\leq 10^6$/ml) in cold 1 M NaHCO$_3$, pH 11; vortex to mix and incubate for 30 min on ice.

2. Centrifuge the extract at 10 000 × g for 10 min. Centrifuge the supernatant in an SW40 or SW40.1 rotor at 150 000 × g (28 000 rev/min) for 1.5 h at 4 °C to pellet the membranes.

3. Wash the pellet twice with PBS containing no divalent cation, each time taking care to wash down the sides of the tube to remove remaining base solution.

   • Omission of Mg^{2+} is important, since it may cause vesiculation of the membrane sheets normally obtained by this method, interfering with the effectiveness of subsequent washes.

4. In initial membrane preparations, monitor the washed membranes by testing the pH of the final resuspended pellet and by running SDS-PAGE on untreated cells and membranes from an equivalent number of cells to determine the effectiveness of extraction. If actin is still present, wash a third time in PBS.

   • Coomassie Blue staining of proteins should be greatly decreased, but lectin blots (Con A is a good choice) should show numerous glycosylated species.

Binding of F-actin

1. Incubate actin alone (50 µg/ml), membranes alone, and membranes with actin in polymerizing buffer (100 mM KCl, 1 mM MgCl$_2$, 50 µM ATP, 10 mM imidazole, pH 7.5) for 2 h at 4 °C, or for 1 h at either room temperature or 37 °C (some interactions appear to be temperature-sensitive and do not occur at low temperatures).

   • Note: Tris is frequently used in actin-polymerizing buffers. However, its effective pH is very temperature-sensitive; moreover, Tris is a reactive amine capable of modifying proteins, possibly altering their function. For an excellent review of buffers, see ref. 86.

2. Extract the membranes for 10 min on ice with 0.2 per cent Triton X-100 or 1.0 per cent octylglucoside in 100 mM KCl, 2 mM MgCl$_2$, 1 mM EGTA, 1 mM PMSF, 10 mM imidazole, pH 7.5.

3. Centrifuge the extract at 50 000 × g for 1.5 h. Analyse the membrane, actin, and membrane + actin extract pellets and dialysed supernatants by SDS-PAGE and lectin blot.

Optimally, the centrifugation conditions should not pellet either the membrane glycoproteins or the actin in the absence of the other. Centrifugation conditions may need to be altered, depending on the size of the membranes and microfilaments produced in a given experiment.

Several considerations are very important in the preparation of base-stripped membranes, which have been made using either sodium hydroxide (85, 87) or sodium bicarbonate (32, 87). When functional studies are to be performed using the membranes, it should be remembered that high pH conditions, especially in sodium hydroxide, may irreversibly denature the membrane protein. Since sodium hydroxide treatment cleaves peptide bonds as well as some glycosyl bonds, use of sodium bicarbonate at high concentration is preferable. Even then all operations should be performed on ice and the time of incubation in base minimized. In our hands it has not been necessary to homogenize the cells or to wash to membranes at high pH in order to remove remaining actin and other known peripheral proteins.

The 'stripped' membrane preparation, when washed thoroughly, contains no actin or other known peripheral proteins. The membranes, either with or without extraction, may be used to assay for the binding of other purified peripheral or cytoskeletal proteins, using biochemical, immunological, or electron microscopy approaches.

2.3.3 Binding of purified proteins to artificial membranes

In numerous instances it has been found that a purified membrane skeletal protein binds to phospholipid vesicles. Some skeletal proteins, including synapsin (88), intermediate filament subunit proteins (89) and vinculin (90) interact with both phospholipid head groups and the hydrophobic fatty acyl tails. For studying the protein-lipid hydrophobic interactions special photo-affinity probes, variously labelled, are being developed. Binding to a specific phospholipid head group is approached by binding the skeletal protein to synthetic large unilamellar vesicles (LUVS). Methods for the preparation of LUVS and reconstitution of synapsin with the vesicles are given in *Protocols 9* and *10*.

Protocol 9. Preparation of mixed phospholipid large unilamellar phospholipid vesicles

1. Dissolve lipids (phosphatidylcholine/phosphatidylethanolamine/phosphatidylserine/phosphatidylinositol/cholesterol, 40:32:12:5:10) (w/w) in a glass test tube in chloroform at a concentration of 10 mg/ml. Add a trace amount of [^{14}C]phosphatidylcholine.

2. Dry the lipids to a thin film in the cold under a gentle stream of nitrogen. Remove traces of solvent by drying the lipid under vacuum for at least 1 h.

3. Dissolve the lipid film in 0.5–1.0 ml of reconstitution buffer (137 mM NaCl, 3 mM sodium azide, 25 mM Tris-HCl, pH 7.4, containing 3 per cent (w/v)) octylglucoside. Keep the final detergent: lipid ratio > 10:1.

4. Stir the mixture gently under nitrogen for 20 min at room temperature.

5. To obtain unilamellar vesicles, remove the detergent by extensive dialysis against cold nitrogen-saturated reconstitution buffer or by gel filtration on a 1.5 × 20 cm Sephadex G-50 column equilibrated with the same buffer. After gel filtration, pool the radioactivity/turbidity fractions and concentrate by vacuum dialysis.

6. Dialyse the vesicle sample against 300 mM glycine, 3 mM sodium azide, 5 mM Hepes, pH 7.4. Keep the vesicles on ice; use within 2–3 days.

Protocol 10. Binding of synapsin I to phospholipid vesicles

1. Remove aggregates from the purified synapsin I preparation by centrifugation at 200 000 × g for 15 min.

   - Note: Many proteins which associate with membranes aggregate after purification, particularly after detergent has been removed, and require this step prior to binding to vesicles.

2. Resuspend the vesicle sample (15–25 µg of lipid/assay) in assay buffer (250 mM glycine, 30 mM sucrose, 30 mM NaCl, 0.22 mM azide, 20 µM EDTA, 5 mM Tris-HCl, 4 mM Hepes, pH 7.4, containing 100 µg/ml BSA). Aliquot into several plastic tubes for the binding assay.

3. Add aliquots of synapsin (to give 5–300 nM), using pre-coated pipette tips, and incubate on ice for 60 min.

4. Separate bound from free synapsin by layering each sample on top of 100 µl of 5 per cent (w/v) sucrose in assay buffer and centrifuging at 42 000 rev/min for 30 min in a Beckman Ti42.2 rotor.

5. Aspirate the supernates carefully and dissolve the pellets in 40 µl of spot buffer (10 mM NaCl, 10 mM NaPO$_4$, pH 7.4, containing 1.4 per cent Triton X-100). Take care not to allow the pellets to touch the test-tube walls above the sucrose cushion, to avoid contamination with free synapsin adhering to the walls.

6. Spot 5-20 µl aliquots on to a nitrocellulose sheet and quantify amounts of synapsin by dot immunoblotting.

This protocol uses a mixture of lipids representing the lipid composition of the synaptic vesicle membrane, previously characterized by extraction and thin layer chromatography (91). Experiments using a single phospholipid are more frequently done, particularly when determining the phospholipid specificity of binding. More extensive protocols for extraction, labelling, analysis and reconstitution of membrane lipids may be found in ref. 12 (Chapter 4).

Such specific interactions of membrane skeletal proteins with phospholipid head groups may suggest a regulatory role for those proteins in cell function. The interactions among the phosphoinositides, calcium ions, microfilament-severing proteins and G-actin-binding proteins may be particularly relevant to membrane signalling, cell shape, and metabolism (92). A case in point is the recent observation of the modulation of phosphoinositide interconversion by specific binding of PI-4, 5-bisphosphate to profilin, the G-actin-binding protein. Sequestration of that lipid resulted in an inhibition of phospho-inositide hydrolysis by unphosphorylated phospholipase Cγ_1, which was reversed by phosphorylation of the enzyme by activated EGF receptor tyrosine kinase (93). A similar type of modulation by sequestration of substrate has been reported for the calcium-sensitive phospholipid- and microfilament-binding protein calpactin, which inhibits phospholipase A2 activity by binding to acidic phospholipids (94, 95). Equally complex regulatory phenomena at the membrane–cytoskeleton interface may occur with other membrane skeleton proteins.

3. Conclusions and prospects

Freeze-etch electron microscopy (Chapter 2), in conjunction with immuno-chemical labelling methods, is providing greater insights into the proteins localized at the membrane–cytoskeleton interface. Yet the limited resolution of electron microscopy does not allow actual demonstration of the interactions between proteins. Molecular biological approaches are making great strides in the elucidation of the structures of numerous membrane and cytoskeletal proteins, and are invaluable for delineating domains of binding and for specific amino acid residues critical in binding. Careful mapping studies done with monoclonal antibodies (Chapter 4) are demonstrating binding sites on proteins for other proteins as well as small ligands. Yet the initial demonstration of binding between two proteins, as well as any factors which might modulate these interactions, is the critical first step in such character-izations. More simple and diagnostic methods are needed for the detection and preliminary characterization of previously unidentified complexes of membrane and cytoskeleton proteins. The need for improved methods for examining weak or transient but potentially physiologically relevant inter-actions of cytoskeletal proteins with the membrane is greater now than ever, given the growing recognition of the importance of such interactions in transmembrane signalling and subsequent alteration of cellular function.

Another complicating factor in studying membrane–cytoskeleton inter-actions is the potential complexity of possible modulating factors. A case in point is the growing number of calcium dependent phospholipid-binding proteins, many having known regulatory functions. Some of these have been shown to bind to cytoskeletal proteins as well as to membranes, and their interaction with the cytoskeleton may actually be modulated by interaction with lipid. Adding to the complexity of the interactions are possible requirements for ATP or other cofactors, making buffer composition a crucial element in observing these interactions. These considerations are not addressed by microscopic, immunologic, or molecular biological approaches, but must be approached by biochemical means.

Acknowledgement

I thank Dr K. P. Campbell for the photographs in *Figure 2*. Original work was supported by NIH grant GM33795.

References

1. Carraway, K. L. and Carraway, C. A. C. (1989). *Biochim. Biophys. Acta*, **988**, 147.
2. Jacobson, B. S. (1983). *Tissue and Cell*, **15**, 829.
3. Geiger, B. (1983). *Biochim. Biophys. Acta*, **737**, 305.
4. Carraway, K. L. and Carraway, C. A. C. (1984). *Bioessays*, **1**, 55.
5. Bennett, V. (1989). *Biochim. Biophys. Acta.*, **989**, 107.
6. Mooseker, M. (1985). *Ann. Rev. Cell Biol.*, **1**, 209.
7. Buck, C. A. and Horwitz, A. F. (1987). *Ann. Rev. Cell Biol.*, **3**, 179.
8. Burridge, K. and Fath, C. (1989). *Bioessays*, **10**, 104.
9. Fox, J. E. B. and Boyles, J. K. (1988). *Bioessays*, **8**, 14.
10. Luna, E. J., Wuestehube, L. J., Ingalls, H. M., and Chia, C. P. (1990). *Adv. Cell Biol.*, **3**, 1.
11. Jones, J. C. R. and Green, K. J. (1991). *Current Opin. Cell Biol.*, **3**, 127.
12. Findley, J. B. C. and Evans, W. H., ed. (1987). *Biological membranes: a practical approach*. IRL, Oxford.
13. Nicolson, G. and Singer, S. J. (1972). *Science*, **175**, 720.
14. Singer, S. J. (1990). *Ann. Rev. Cell Biol.*, **6**, 247.
15. Towler, D. A., Gordon, J. I., Adams, S. P., and Glaser, L. (1988). *Ann. Rev. Biochem.*, **57**, 69.
16. Schultz, A. M., Henderson, L. E., and Oroszlan, S. (1988). *Ann. Rev. Cell Biol.*, **4**, 611.
17. Ferguson, M. A. J. and Williams, A. F. (1988). *Ann. Rev. Biochem.*, **57**, 285.
18. Burn, P. (1988). *Trends Biochem. Sci.*, **13**, 79.
19. Carraway, K. L. (1975). *Biochim. Biophys. Acta Reviews on Biomembranes*, **415**, 379.
20. Varki, A. (1991). *FASEB J.*, **5**, 226.
21. Glossman, H. and Neville, D. M. (1971). *J. Biol. Chem.*, **246**, 6339.

22. Steck, T. L. and Dawson, G. (1974). *J. Biol. Chem.*, **249**, 2135.
23. Burridge, K. (1976). *Proc. Natl. Acad. Sci. USA*, **73**, 4457.
24. Bollag, D. M. and Edelstein, S. J. (1990). *Protein methods*. Wiley-Liss, New York.
25. Carraway, K. L. and Carraway, C. A. C. (1982). Isolation and characterization of plasma membranes. In *Antibody as a tool: the application of immunochemistry* (eds J. J. Marchalonis and G. W. Warr), p. 509. Wiley, Chichester.
26. Carraway, K. L., Huggins, J. W., Cerra, R. F., Yeltman, D. R., and Carraway, C. A. C. (1980). *Nature*, **285**, 508.
27. Carraway, C. A. C., Cerra, R. F., Bell, P. B., and Carraway, K. L. (1982). *Biochim. Biophys. Acta*, **719**, 126.
28. Hale, J. E. and Wuthier, R. E. (1987). *J. Biol. Chem.*, **262**, 1916.
29. Carraway, C. A. C., Jung, G., and Carraway, K. L. (1983). *Proc. Natl. Acad. Sci. USA*, **80**, 430.
30. Carraway, C. A. C. and Weiss, M. (1985). *Exp. Cell Res.*, **161**, 150.
31. Carraway, C. A. C., Sindler, C., and Weiss, M. (1986). *Biochim. Biophys. Acta*, **885**, 68.
32. Carraway, C. A. C., Hua, F., Ye, X., Juang, S.-H., Liu, Y., Carvajal, M. E., and Carraway, K. L. (1991). *J. Biol. Chem.*, **266**, 16238.
33. Osborn, M. and Weber, K. (1977). *Cell*, **12**, 561.
34. Lis, H. and Sharon, N. (1986). *Ann. Rev. Biochem.*, **55**, 35.
35. Clarke, A. E. and Hoggart, R. M. (1982). The use of lectins in the study of glycoproteins. In *Antibody as a tool: the application of immunochemistry* (eds J. J. Marchalonis and G. W. Warr), p. 347. Wiley, Chichester.
36. Sharon, N. and Lis, H. (1989). *Lectins*. Chapman and Hall, London.
37. Walker, J. M. (ed.) (1984). *Methods in molecular biology*, Vol. 1, *Proteins*. Humana Press, Clifton, NJ.
38. *Protein Blotting Protocols for the Immobilon-P Transfer Membrane*. Technical bulletin, Millipore Intertech, Bedford, MA.
39. Ervasti, J. M., Ohlendieck, K., Kahl, S. D., Gaver, M. G., and Campbell, K. P. (1990). *Nature*, **345**, 315.
40. Vanderpuye, O., Carraway, C. A. C., and Carraway, K. L. (1988). *Exp. Cell Res.*, **178**, 211.
41. Clegg, J. C. S. (1982). *Anal. Biochem.*, **127**, 389.
42. Gershoni, J. M. and Palade, G. E. (1982). *Anal. Biochem.*, **124**, 396.
43. Glass, W. F., Briggs, R. C., and Hnilica, L. A. (1981). *Anal. Biochem.*, **115**, 219.
44. Robinson, P. J., Bull, F. G., Anderton, B. H., and Roitt, I. M. (1975). *FEBS Lett.*, **58**, 330.
45. Haselbeck, A., Schickaneder, E., van der Eltz, H., and Hosel, W. (1990). *Anal. Biochem.*, **191**, 25.
46. Olden, K., Bernard, B. A., Humphries, M. J., Yeo, T.-K., White, S. L., Newton, S. A., Bauer, H. C., and Parent, J. B. (1985). *Trends Biochem. Sci.*, **10**, 78.
47. Krag, S. S. (1985). *Current Topics in Membranes and Transport*, **24**, 181.
48. West, C. M. (1986). *Molec. Cell. Biochem.*, **72**, 3.
49. Brandley, B. K., Swiedler, S. J., and Robbins, P. W. (1990). *Cell*, **63**, 861.
50. Cuatrecasas, P. and Hollenberg, M. D. (1976). *Adv. Protein Chem.*, **30**, 251.

51. Fox, J. E. B. (1985). *J. Biol. Chem.*, **260**, 11970.
52. Ezell, R. M., Kenney, D. M., Egan, S., Stossel, T. P., and Hartwig, J. H. (1988). *J. Biol. Chem.*, **263**, 13303.
53. Painter, R. G., Gaarde, W., and Ginsberg, M. H. (1985). *J. Cell. Biochem.*, **27**, 277.
54. Koch, G. L. E. and Smith, M. J. (1978). *Nature*, **273**, 274.
55. Campbell, K. and Kahl, S. D. (1989). *Nature*, **338**, 259.
56. Bennett, V. (1985). In *Cell membranes: methods and reviews*, Vol. 2 (eds E. Elson, W. Frazier, and L. Glaser), p. 149. Plenum Press, New York.
57. Otey, C. A., Pavalko, F. M., and Burridge, K. (1990). *J. Cell Biol.*, **111**, 721.
58. Argraves, W. S., Dickerson, K., Burgess, W. H., and Ruoslahti, E. (1989). *Cell*, **58**, 623.
59. Pardee, J. D. and Spudich, J. A. (1982). In *Methods in cell biology*, Vol. 24, (ed. L. Wilson), p. 271. Academic Press, New York.
60. Cooper, J. A. and Pollard, T. D. (1982). *Meth. Enzymol.*, **85**, 182.
61. Hartwig, J. W. and Kwiatkowski, D. J. (1991). *Curr. Opin. Cell Biol.*, **3**, 87–97.
62. Kremer, L., Dominguez, J. E., and Avila, J. (1988). *Anal. Biochem.*, **175**, 91.
63. Snabes, M. C., Boyd, A. E. III, and Bryan, J. (1981). *J. Cell Biol.*, **90**, 809.
64. Wang, Y.-L. and Taylor, D. L. (1980). *J. Histochem. Cytochem.*, **28**, 1198.
65. Kouyama, T. and Mihashi, K. (1981). *Eur. J. Biochem.*, **114**, 33.
66. Dos Remedios, C. G. and Cooke, R. (1984). *Biochim. Biophys. Acta*, **788**, 193.
67. Okabe, S. and Hirokawa, N. (1989). *J. Cell Biol.*, **109**, 1581.
68. Snabes, M. C., Boyd, A. E., III, and Bryan, J. (1981). *J. Cell Biol.*, **90**, 809.
69. Safer, D. (1989). *Anal. Biochem.*, **178**, 32.
70. Grandmont-Leblanc, A. and Gruda, J. (1977). *Can. J. Biochem.*, **55**, 949.
71. Winstanley, M. A., Trayer, H. R., and Trayer, I. P. (1977). *FEBS Lett.*, **77**, 239.
72. Ohara, O., Takahashi, S., Ooi, T., and Fujiyoshi, Y. (1982). *J. Biochem.*, **91**, 1999.
73. Luna, E. J., Wang, Y.-L., Voss, E. W., Jr., Branton, D., and Taylor, D. L. (1982). *J. Biol. Chem.*, **257**, 13095.
74. Miller, K. G. and Alberts, B. M. (1989). *Proc. Natl. Acad. Sci. USA*, **86**, 4808.
75. Wuestehube, L. J., Speicher, D. W., Shariff, A., and Luna, E. J. (1991). *Methods. Enzymol.*, **196**, 47.
76. Luna, E. J., Goodloe-Holland, C. M., and Ingalls, H. M. (1984). *J. Cell Biol.*, **99**, 58.
77. Luna, E. J., Wuestehube, L. J., Chia, C. P., Shariff, A., Hitt, A. L., and Ingalls, H. M. (1990). *Devel. Genetics*, **11**, 354.
78. Wuestehube, L. J. and Luna, E. J. (1987). *J. Cell Biol.*, **105**, 1741.
79. Coleman, T. R., Harris, A. S., Mische, S. M., Mooseker, M., and Morrow, J. S. (1987). *J. Cell Biol.*, **104**, 519.
80. Spudich, J. A. and Watt, S. (1971). *J. Biol. Chem.*, **246**, 4866.
81. Burridge, K. and Mangeat, P. (1984). *Nature*, **308**, 744.
82. Hummel, J. P. and Dreyer, W. J. (1962). *Biochim. Biophys. Acta*, **63**, 530.
83. Horwitz, A., Duggan, K., Buck, C., Beckerle, M. C., and Burridge, K. (1986). *Nature*, **320**, 531.
84. Horwitz, A., Duggan, K. Greggs, R., Decker, C., and Buck, C. (1985). *J. Cell Biol.*, **101**, 2134.

85. Shariff, A. and Luna, E. (1990). *J. Cell Biol.*, **110**, 681.
86. *Buffers*. Technical bulletin, Calbiochem, Division of Hoechst International Corp., San Diego, CA.
87. Hubbard, A. L., Wall, D. A., and Ma, A. (1983). *J. Cell Biol.*, **96**, 217.
88. Benfenati, F., Greengard, P., Brunner, J., and Bähler, M. (1989). *J. Cell Biol.*, **108**, 1851.
89. Perides, G., Harter, C., and Traub, P. (1987). *J. Biol. Chem.*, **262**, 13742.
90. Niggli, V., Dimitrov, D. P., Brunner, J., and Burger, M. M. (1986). *J. Biol. Chem.*, **261**, 6912.
91. Westhead, E. W. (1987). *Ann. New York Acad. Sci.*, **493**, 92.
92. Carraway, K. L. (1990). *Bioessays*, **12**, 90.
93. Goldschmidt-Clermont, P. J., Kim, J. W., Machesky, L. M., Rhee, S. G., and Pollard, T. D. (1991). *Nature*, **261**, 1231.
94. Davidson, F. F., Dennis, E. A., Powell, M., and Glenney, J. R., Jr. (1987). *J. Biol. Chem.*, **262**, 1698.
95. Haigler, H. T., Schlaepfer, D. D., and Burgess, W. H. (1987). *J. Biol. Chem.*, **262**, 6921.

7

Analysis of microtubule dynamics
in *vitro*

ROBLEY C. WILLIAMS, Jr

1. Introduction

Microtubules exhibit a spontaneous rapid growth and shortening, a property called *dynamic instability* (1, 2, 3). The causes and consequences of this phenomenon, its origin in molecular interactions, the means by which it is modulated, and its consequences for cellular functioning are all subjects of active current investigation. Although the phenomenon was discovered by electron microscopy (1, 2) and can be studied by biochemical methods that measure the average properties of ensembles of microtubules (4, 5, 6), the observation of individual microtubules by light microscopy is probably most revealing. By following the growth and shortening of single intact microtubules over a period of time (7, 8, 9), one can easily observe details of their dynamics that are virtually inaccessible by other methods.

The description in this chapter is aimed at the molecular biologist or biochemist who wishes to observe microtubule dynamics *in vitro*, adding or subtracting molecular components of the system and interpreting observations at the intermediate level of structural resolution provided by light microscopy. The apparatus and technique described are relatively simple and inexpensive. Dynamic instability has also been observed by similar means in the microtubules of living cells. The reader is referred to the appropriate literature for details (10, 11, 12).

Microtubules can be visualized in darkfield or in differential-interference-contrast (DIC or Nomarski) microscopy. Advantages and limitations apply to each technique, and the DIC method is described here. In addition to early descriptions of video-enhanced DIC microscopy (13, 14), practical articles (7, 15) and a comprehensive book (16) can be consulted for details and variations not covered here.

2. How to do it: a brief overview

One prepares a solution containing tubulin at a concentration of 5 to 30 μM in an appropriate GTP-containing buffer. Particles that can adhere to the glass surfaces and which serve as nuclei for the formation of microtubules are added to the solution; a small amount if then placed between slide and cover slip. Microtubules are caused to assemble by raising the temperature to around 37 °C and are observed by means of a DIC microscope equipped with a video camera. The structures, not visible when one simply looks through the eyepieces, are detected by making use of a frame buffer to subtract an image of the background from the image of the scene and then electronically raising the contrast of the resulting 'difference' image. The result is viewed on a monitor and recorded on videotape for later measurement.

3. Equipment

3.1 Microscopy

3.1.1 Strategic considerations

A single microtubule has a small diameter and a refractive index close to that of water. It makes only a small perturbation of the optical wavefront, and consequently produces an image of very low contrast in DIC microscopy. Microscopic apparatus for the observation of single microtubules therefore concentrates on enhancing contrast to make the structures visible. Contrast enhancement has the undesired side-effect of accentuating unevenness in the optical background. Spatial variation in the intensity of the illuminating light and in the sensitivity of the video camera, unevenness of the glass surfaces of the specimen chamber and dust in the optical system all contribute to this background. Subtracting an image of the defocused specimen (a 'background image') from the primary video image partially corrects the problem, yielding a relatively uniform corrected image whose contrast can be successfully increased.

3.1.2 Microscope and video camera

A typical apparatus is illustrated in *Figure 1*, and its components are discussed in *Table 1*. We use a Zeiss Axiovert 35 microscope (any other sturdy microscope body would work), mounted on a marble balance table as a precaution against vibration transmitted from the floor. The Zeiss DIC optical system works well in this application. (Experience of other laboratories has indicated that the systems of some other manufacturers do not allow visualization of microtubules at all, possibly because the beam-splitting Wollaston prism that they employ produces too great a lateral separation of the two orthogonally polarized rays that traverse the specimen. A microscope being purchased for the purpose of observing microtubules should be tested before money is spent.)

Both the contrast of microtubules and the signal/noise ratio of the camera increase with the intensity of illumination (15). The mercury arc is used because it provides a greater luminous density over a narrow band of visible

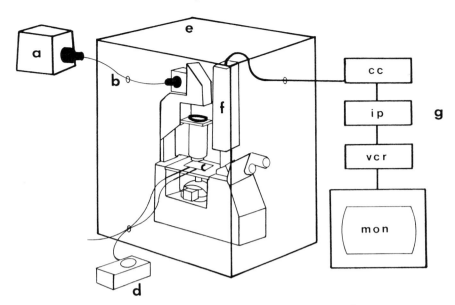

Figure 1. Schematic diagram of microscope and electronics. *a*, lamphouse; *b*, scrambler; *c*, flow cell and attached tubing; *d*, peristaltic pump; *e*, temperature-controlled enclosure; *f*, video camera; *g*, electronics (*cc*; camera control, *ip*; image processor, *vcr*; videotape recorder, *mon*, monitor).

wavelengths than does a xenon arc. A disadvantage of the mercury arc is that it produces highly uneven illumination. This disadvantage is overcome by the use of a fibre-optic scrambler (17, 18), a recently invented device that subjects the illuminating light to multiple internal reflections within a single quartz fibre and thereby increases the uniformity of illumination of the condenser aperture. The result is excellent visibility of the microtubules. Since colour correction and absolute flatness of field are not required, the Plan-Neofluar objective produces good results at low cost.

The video camera is solidly coupled to the microscope via a transfer lens and C-mount adapter (Carl Zeiss, Inc) configured to provide nearly unit magnification. The edges of the visible field of the microscope extend slightly beyond the edges of the camera's target. Although our camera has variable black level, sensitivity, and contrast, these features are used only during adjustment of the optics. During data-gathering the 'automatic' mode is quite satisfactory.

3.1.3 Image processing and display

The minimal requirements for an image-processing computer are that it be able to average multiple successive frames, to store one video frame as a background, to subtract that stored frame in real time from the incoming

Table 1. Equipment for video-enhanced DIC microscopy

Component	Comments
DIC microscope	Zeiss equipment performs well in this application. Desired qualities are mechanical stability, a convenient port for the video camera, and ease of alignment
illuminating system	An Osram type HBO 100 W/2 mercury arc, a 546-nm interference filter, and a condenser of 1.4 NA provide sufficient (but not excess) illumination. The 'scrambler' (Technical Video, Ltd) is an optional device that considerably increases visibility of microtubules in this optical system by producing even illumination of the condenser diaphragm
objective lens	Zeiss Plan-Neofluar 100 ×, NA 1.3
specimen chamber	A simple chamber (*Figure 3A*) or a flow-cell (*Figure 3B*) (available from Ken-Tex Machine Shop) can be employed. Temperature control by means of an air-bath, air-curtain, or thermoelectrically regulated stage is desirable
temperature box	Plexiglas. Swinging front door allows major adjustments. Curtained handports left and right allow specimen exchange and focusing. Thermostat (Thermistemp, Model 63RC, fan (an electronics cooling fan works well), and heater maintain temperature.
video camera	Newvicon (e.g. Dage Model 67, Dage-MTI, Inc. or CCD camera. Cameras that have an automatic control of output signal level work well. The additional availability of a manual setting for this function (as on the Dage 67) is useful when adjusting the DIC system because it allows the absolute intensity on the video screen to be coupled to the absolute intensity at the camera's target
image processor	Device to provide background subtraction and subsequent manipulation of image contrast, e.g. Hamamatsu Argus 10 (available from Hamamatsu Photonic Systems Corp), or the more elaborate Image I system (Universal Imaging).
videotape recorder	Super-VHS-format recorder, such as the Mitsubishi BV-1000, provides adequate resolution at moderate price. Important features are the capacity to advance the tape one frame at a time and the 'jog-shuttle', which allows forward-and-back examination of the tape.
Monitor	Super-VHS compatible monitor, e.g. Sony PVM 2530 or Sony KV20TS30. A useful feature is the ability to DC-couple the monitor to the signal (by bypassing the DC-restoration function), permitting the useful display of absolute intensity on the monitor.

video image, and to manipulate in real time the contrast of the corrected video image. The relatively inexpensive Hamamatsu Argus 10 carries out these functions digitally and allows facile adjustment of the contrast transfer function. A variety of other systems are commercially available, many designed to allow (at higher cost) measurements of intensities and morphological properties of the image in addition to the minimal functions described above.

Videotape is the least expensive way of storing large numbers (hours) of video frames. Super-VHS recorders equipped with a 'jog-shuttle' feature provide good resolution (about 425 lines in the horizontal direction) and workable access to individual single frames. Optical memory disk recorders (OMDRs) allow, at much higher cost, rapid computer-controlled access to particular single frames.

Good quality Super-VHS monitors are available on the consumer market. Ours has a 64-cm (25-inch) screen. When combined with the 100× objective, a 2.5× intermediate lens and the particular camera (standard target dimensions) with its transfer lens, it provides an overall magnification of approximately 10 000×. Thus, 1 μm in the sample appears as 1 cm on the screen.

The electronics associated with the Super-VHS format separate the colour-generating (C, or chrominance) signal from the intensity-generating (Y, or intensity or luminance) signal, assigning each a separate wire. Since there is no intrinsic chromatic information in the microscopic image, and since both the camera and the image processor are monochrome devices, it is best to disable the colour-generating feature of recorder and monitor, making an effective black-and-white system. This is easily done by building a cable that connects the 'video' connector of the image processor to the 'Y' lead of the S-VHS system, as shown in *Figure 2*.

Temperature control of the sample is necessary. We have chosen to house the microscope and camera in a plexiglas box and to employ a fan and a thermostatically-regulated heater to maintain them constantly at 37 °C. A first reason for doing this is that the specimen is necessarily connected by immersion oil to both condenser and objective, and could be substantially cooled by these relatively massive thermally-conductive elements. A second reason is the observation (E. D. Salmon, personal communication) that continued thermal cycling of the objective lens introduces stress, which gradually renders the lens birefringent and thus reduces contrast in the DIC system.

3.1.4 Chambers and flow cells

Simple chambers for the observation of microtubules can be made by applying a drop of sample (5 μl minimum volume) to a clean slide, placing a

Figure 2. S-VHS plug, viewed with the pins pointing at the observer. To make a video-to-S-VHS connector, cut an S-VHS plug from the end of an interconnecting cable, leaving a few cm of wire attached. Connect pins 1 and 2 (the C-signal) together. Connect pin 3 (the Y-ground) to the outer shield (ground) of the coaxial video cable. Connect pin 4 (the Y-signal) to the central (signal) wire of the coaxial cable.

clean 22 × 22 mm #0 cover slip on top of the drop and quickly sealing with valap (a warmed mixture of equal quantities of vaseline, lanolin, and paraffin). We have found that substantial adsorption of tubulin (and other proteins) to glass can significantly reduce the concentration present in such a chamber (cf. ref. 7). To lower the surface/volume ratio, a spacer of parafilm can be added. The cell shown in *Figure 3A* has proven to be easy to fill and to seal.

A flow cell is useful for changing the solution surrounding microtubules that have already formed, and for a variety of other experiments. That shown in *Figure 3B* is a modification of a published design (19). A small syringe pump is used to exchange solutions.

3.2 Measurement

It is convenient to separate the measurement process from observation by measuring videotaped images. The fundamental requirements for a measurement apparatus are that it be able to superimpose on the continuous (30 frames/s) stream of video frames a cursor that can be moved by the investigator to mark the end of the microtubule, and that it be capable of recording the position of the cursor (by transfer to a computer memory) when the investigator pushes a button. Many commercially available image-analysis systems will do this. We have adopted as a low-cost solution a mouse-driven board (Video Van Gogh Board, Tekmatic Systems) that plugs into an IBM AT computer and a program that operates it. (The program RTM, written in Turbo Pascal for the IBM AT by Stephen Magers and E. D. Salmon, Department of Biology, University of North Carolina, Chapel Hill, NC27599–3280, USA) provides a means of making mouse-based length measurements in combination with the Video Van Gogh board.) The board divides the screen into a 256 × 256 array of positions. The smallest increment of distance that can be measured in the sample is therefore about 0.2 μm, just below the resolution of the microscope. To measure slow events, the investigator simply follows the end of the microtubule with the cursor, pushing the button whenever a measurement is desired. To measure rapid events, it is often useful to advance the videotape recorder frame by frame, measuring the static image at appropriate intervals.

Calibration of magnification should be carefully carried out, in both the horizontal and the vertical directions, by the use of a stage micrometer. Different locations on the screen should be checked as a precaution against geometric distortions that may be present in the camera, leading to different effective magnifications at different places. Once measurements are made, it is useful to be able to represent them graphically, either as plots of length versus time or in some other form. Most measurement programs have such a feature.

Images for publication can be made by photographing the screen. Shutter

156

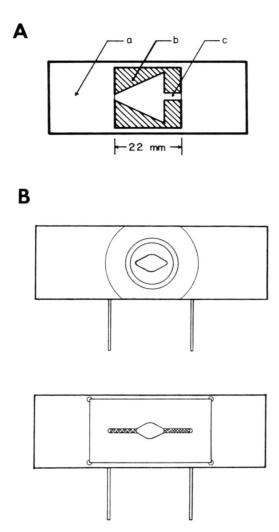

Figure 3. *A*, Schematic diagram of a cell made from a slide (*a*) and cover slip (*c*), with a parafilm spacer (*b*, hatched). To make the cell, cut the spacer, sandwich it between a clean slide and cover slip, and warm the sandwich slowly on a hotplate until the parafilm begins to melt. Seal with a bead of valap, leaving the holes open for filling. This shape of spacer allows easy filling of the cell. *B*, Schematic diagram of flow cell modified from reference 19. The side closest to the objective lens is at top. A 10 mm round cover slip on that side is separated from a 22 × 40 mm rectangular cover slip on the side facing the condenser. The chief modifications from the published design are in the shape of the chamber, which is roughly diamond-shaped rather than rectangular, and in the filling channels (cross-hatched), which replace the filling troughs in the original. These modifications make the cell substantially less likely than the original design to trap bubbles.

openings longer than 1/15 second are required for a clear image. Many image processors have a routine that will average a specified number of frames, then stop and display the averaged image. The average of 4 to 16 frames is often markedly clearer than a single frame.

4. Experimental methods

4.1 Preparation of tubulin

Tubulin can be made by any of a number of routes (20, 21). The method in *Protocol 1* yields approximately 400 mg of essentially pure tubulin.

Protocol 1. Preparation of tubulin (modified from ref. 20)

1. Collect 4−6 bovine brains (1.6−2.4 kg). Quickly chill by placing them in a plastic bag surrounded by ice. Working in a cold room, remove meninges and superficial blood vessels and coarsely mince the tissue with scissors.

2. Homogenize the tissue, 100 g at a time, with 75 ml of PM-4M buffer (0.1 M Pipes, 2 mM EGTA, 1 mM $MgSO_4$, 2 mM DTE, 4 M glycerol, adjusted to pH 6.9 with NaOH) in a blade-type homogenizer. If a Sorvall Omni Mixer is used, allow 50 s at speed 3 and 10 s at speed 10.

3. Centrifuge the homogenate for 15 min at 6500 × g (8000 rev/min in a Sorvall type GSA rotor). Decant the supernatant and centrifuge it for 75 min at 96 000 × g (35 000 rev/min in a Beckman Type 35 rotor). Decant the supernatant, measuring its volume, V_1 (approximately 700 ml). Add sufficient GTP to make 1 mM.

4. Add sufficient ATP to bring to 2.5 mM and pour the supernatant into centrifuge tubes. Balance the tubes for centrifugation and then incubate for 30−45 min in a water bath at 34 °C. During this incubation, microtubule assembly should take place, and should be indicated by growing opalescence and increased viscosity of the solutions.

5. Centrifuge at 30−34 °C in a prewarmed rotor for 60 min at 96 000 × g to pellet microtubules. Decant the supernatant and save the pellets.

6. Resuspend the pellets in a total of 0.25 V_1 of ice-cold PM buffer (0.1 M Pipes, 2 mM EGTA, 1 mM $MgSO_4$, 2 mM DTE, 1 mM GTP, adjusted to pH 6.9 with NaOH), with the aid of a rubber policeman. Decant the coarsely resuspended material into a 60-ml Dounce homogenizer and subject to about five gentle strokes of the pestle to produce a uniform suspension. Incubate 30 min on ice to depolymerize microtubules.

7. Pour into cold centrifuge tubes. Centrifuge at 4 °C for 60 min at 96 000 × g. Save the supernatant, measuring its volume, V_2. Discard the pellets.

8. To the pooled supernatant add an equal volume of PM-8M buffer (0.1 M Pipes, 2 mM EGTA, 1 mM $MgSO_4$, 2 mM DTE 8 M glycerol, adjusted to pH 6.9 with NaOH).

9. Carry out a second cycle of polymerization, centrifugation, depolymerization, and centrifugation by repeating steps 4–8, replacing V_1 by V_2 in step 6. The resulting material, referred to as twice-cycled microtubule protein, can be stored frozen at −70 °C if necessary.

10. Prepare a 2 × 45 cm column of Whatman P-11 phosphocellulose by carefully precycling according to the manufacturer's instructions and equilibrating with column buffer (0.1 M Pipes, 2 mM EGTA, 1 mM $MgSO_4$, 2 mM DTE, 0.1 mM GTP, adjusted to pH 6.9 with NaOH). Check the pH and conductivity of the eluant to make sure they are the same as those of the buffer flowing into the column.

11. Apply the entire solution from step 9 to the column at 4 °C, and elute the tubulin with column buffer, collecting fractions of approximately 2 ml. Soon after the fractions emerge from the column, add to each tube a volume of 0.1 M $MgSO_4$ sufficient to bring the Mg^{2+} concentration to 1 mM. This essential step replaces Mg^{2+} that is removed from solution by the phosphocellulose (23). (A fraction containing microtubule-associated proteins can be eluted from the column by application of column buffer +0.8 M NaCl).

12. Pool those fractions with concentrations greater than 1 mg/ml. The yield should be approximately 400 mg of tubulin of > 96 per cent purity. (It is wise to examine the product by SDS-PAGE.) Store tubulin in small aliquots frozen quickly by immersion in liquid nitrogen and kept at −70 °C. Eppendorf centrifuge tubes are convenient for the purpose. Tubulin treated in this way must be equilibrated after thawing with the buffer in which it is to be used.

4.2 Preparation of axonemal pieces

Sperm-tail axonemes act as excellent nuclei for microtubules, adhering to glass so that one end of the microtubule is anchored against Brownian motion and bulk flow of buffer. *Protocol 2* gives a method for their preparation from sea-urchin sperm.

Protocol 2. Preparation of axonemal fragments from sea urchin sperm (modified from refs 24 and 7)

1. Begin with 5 ml of packed sperm, previously washed several times in sea water + 0.1 mM EDTA.

2. Wash twice by resuspending in 100 ml of 0.15 M NaCl + 1 mM EDTA and centrifuging 7 min at 3 000 × g.

Protocol 2. *Continued*

3. Resuspend the packed pellet in 50 ml of 20 per cent sucrose in distilled water and homogenize in a large Dounce homogenizer with eight vigorous strokes of the tight-fitting (A) pestle. This should detach heads from tails and effect fragmentation of the membranes.

4. Centrifuge the homogenate at 3000 × *g* (6200 rev/min in a Beckman Type 35 rotor) for 7 min to pellet most of the sperm heads and membrane fragments. Carefully remove the supernatant from the loosely-packed pellet. Put the pellet aside.

5. Centrifuge the supernatant at 27 000 × *g* (18 500 rev/min in a Type 35 rotor) for 15 min. The axonemes should form a thick off-white layer at the top of the stratified pellet. Resuspend this layer in about 30 ml of buffer (0.1 M NaCl, 5 mM Hepes, pH 7, 4 mM $MgSO_4$, 1 mM $CaCl_2$, 1 mM EDTA, 7 mM 2-mercaptoethanol), discarding the dark layer at the bottom and any diffuse yellow layer that may exist. Homogenize with five gentle strokes of the loose-fitting (B) pestle of the Dounce homogenizer.

6. Centrifuge the resuspended axonemes at 12 000 × *g* (12 400 rev/min, Type 35 rotor) for 10 min. Remove and set aside the supernatant.

7. To remove dynein outer arms, gently resuspend the pellet in 30 ml of buffer to which has been added an additional 0.5 M NaCl, discarding any dark material at the bottom. Incubate 30 min on ice.

8. Remove residual sperm heads by layering the preparation on a layer of 80 per cent sucrose and centrifuging at 16 000 × *g* for 10 min. Carefully remove the top and interfacial layers containing the axonemes.

9. By trial-and-error dilution of small aliquots and assay with the microscope, find a concentration that gives a usable number of axonemes per unit area (1–10 per field) when observed at operating magnification.

10. Prepare 5-μl aliquots at a concentration 100-fold greater than this and store frozen at −70 °C. Axonemes treated in this way will keep for at least a year.

4.3 Starting an experiment

Protocols 3 and 4 describe the steps to prepare a specimen for observation by combining tubulin with axonemes and raising the temperature to initiate assembly.

Protocol 3. Initiating microtubule assembly in a sealed chamber

1. Keep all samples on ice until their introduction in to the sample chamber.

2. Add 245 µl of buffer (0.1 M Pipes, 1 mM MgSO₄, 2 mM EGTA, 1 mM DTE, 1 mM GTP, adjusted to pH 6.9 with NaOH) to a 5 µl aliquot of axonemes, prepared as described in *Protocol 2*.

3. Thaw an aliquot of tubulin by immersing its Eppendorf tube in water at room temperature, being careful to remove the tube as the last of the frozen sample melts, so that the temperature remains low. Centrifuge the tube for 10 min in an Eppendorf centrifuge (15 600 × *g*) to pellet any denatured protein. Remove the supernatant and dilute to 3 mg/ml.

4. (a) To fill a chamber without spacers, pipette 2 µl of diluted axonemes and 2 µl of buffer into the bottom of an ice-cold small Eppendorf tube. Add 3 µl of tubulin solution. Pipette the mixture on to a clean cover slip (22 × 22 mm, #0), spreading around with the pipette tip to make a puddle about 4 mm in diameter. Quickly place the cover slip onto the slide and seal with warm valap, using a toothpick to draw the warm liquid around the edges of the cover slip. (b) To fill a chamber made with a spacer of the type shown in *Figure 3A*, pipette 15 µl of diluted axonemes and 15 µl of buffer into a cold Eppendorf tube. Add 30 µl of tubulin. Mix. Without delay, draw the mixture into a pipette. Then, touching the tip of the pipette to the narrow end of the triangular chamber, slowly eject the solution, allowing it to fill the chamber by capillary action. Seal the filling holes with valap.

5. Quickly apply immersion oil, warmed to 37 °C, to the top and bottom of the sealed chamber, and place on the warm microscope stage, bringing the condenser and objective lens into focus.

Protocol 4. Initiating microtubule assembly in a flow cell

1. Keep all samples on ice until their introduction to the sample chamber.

2. Degas all solutions by exposing them briefly to vacuum.

3. Add 245 µl of buffer (0.1 M Pipes, 1 mM MgSO₄, 2 mM EGTA, 1 mM DTE, 1 mM GTP, adjusted to pH 6.9 with NaOH) to a 5-µl aliquot of axonemes, prepared as described in *Protocol 2*.

4. Apply immersion oil to the windows of the clean flow chamber, which has been previously flushed with assembly buffer (0.1 M Pipes, 2 mM EGTA, 1 mM MgSO₄, 2 mM DTE, 1 mM GTP, adjusted to pH 6.9 with NaOH). Place it on the stage of the microscope and bring the inner surface nearest the objective lens into focus. Introduce axonemes by flowing through a volume equal to 1.5 times the total volume of the cell and its associated

Protocol 4. *Continued*

tubing. Allow to stand for 10 min. Flush out axonemes that have not adhered with a flow (2–5 μl/s) of assembly buffer.

5. Prepare about 500 μl of tubulin in assembly buffer (concentration 0.7–3.0 mg/ml), kept on ice. Slowly introduce the protein into the chamber, stopping the flow after several cell-volumes have passed through. To keep the inlet tubing clear of assembled microtubules, one can follow the tubulin solution with about 0.5 tubing volumes of buffer.

6. Observe.

4.4 Making taped observations

4.4.1 Adjustment for maximum contrast and background subtraction

Protocol 5 gives instructions for obtaining a good image, and two examples are shown in *Figure 4*. The technique described in *Protocol 5* for setting the DIC Wollaston prism is quite subjective and is somewhat poorly controlled. Reference 15 describes modifications that can be made to the microscope to allow more reproducible setting of the DIC system.

Protocol 5. Adjustment for a good image

1. With the arc lamp well warmed up, adjust the microscope for Koehler illumination. In some cases it may be necessary to adjust for critical illumination in order to gain sufficient intensity (18).

2. Looking through the eyepieces, or watching the monitor if it and the camera can be made to respond directly to intensity, adjust the second Wollaston prism (the beam recombiner) to produce the darkest possible image. Then adjust *slightly* away from this position. (Practice with an unimportant specimen will give one a feel for the optimal setting.) If the camera is not brightly enough illuminated to produce a satisfactory image, move the Wollaston prism further.

3. Defocus the objective lens, moving the focal plane into the glass of the cover slip. Using the image processor, record the average of several hundred frames (we use 256) of this 'background' image.

4. Refocus the objective to the plane of interest and activate the image processor's background subtraction mode to produce an initial image. Adjust the processor transfer function (usually a lookup table) so that the dimmest input pixels are almost black on output and the brightest input pixels are almost white.

5. If microtubules are not clearly visible, try steps 2–4 again, but with a different adjustment of the DIC prism.

(a)

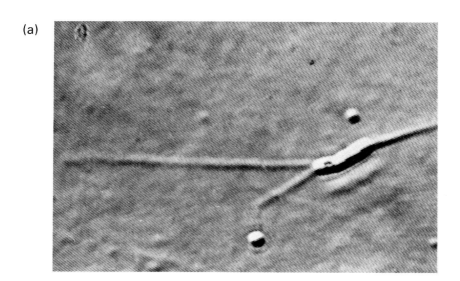

(b)

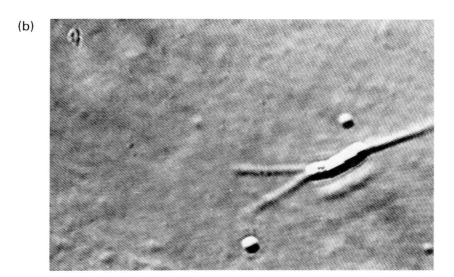

Figure 4. Micrographs of two microtubules growing from the end of a piece of flagellar axoneme. Photograph (b) was taken 15 s after photograph (a). Note that one of the microtubules has shortened appreciably. Note also that in (a) the position of the tip of the longer microtubule is less easily measured than that of the short one. The relative ease of measurement is reversed in (b). Each photograph was taken from a still image that was the average of four consecutive frames (4/30 s).

4.4.2 Focus and background adjustments

Microtubules that grow outward from axonemes tend to lie nearly parallel to the glass surface, making a small angle upwards into the solution. The depth of focus in the microscope described here is small (less than 0.5 μm), so that the whole extent of a long microtubule cannot be kept in focus. It is the position of the tip of the microtubule that must be measured to assess the dynamics of growth of the structure. But when changes in length occur, especially shortening of the microtubule, the focal depth of the tip can change enough to make measurement difficult. A good strategy is to focus on a stretch of microtubule a few μm from the tip. To make video tape that can be measured easily, it is necessary that the operator watch continually and 'trim' the focus occasionally, but not so often that the drifting of the top due to Brownian motion is followed.

When the intensity of the lamp fluctuates slowly and the degree of contrast enhancement is great, it can be necessary to record a fresh background at intervals in order to maintain good image quality. If the focus is carefully restored, this can be accomplished without substantially displacing the image relative to the field of view.

4.4.3 Timing

It is advisable for many purposes to find a good field and watch it continuously for a long period. However, since tubulin is subject to show spontaneous degradation, a sample should not normally be used for more than an hour. The time-and-date generator built into most image processors is useful for timing an experiment on the tape.

4.4.4 Use of the flow cell

The volume required to exchange the contents of the flow cell completely can be determined by passing a dye through the system with a syringe pump, observing the absorbance of the material in the chamber by means of a spectrophotometer. When material is exchanged during an experiment, this volume should always be kept in mind.

4.5 Measurements

Measurements of microtubule length are made by taking the apparent point at which the microtubule joins the axoneme as one end and the tip as the other. Flexibility of the structure is small enough that it can be presumed to be a straight line within an error of only a few per cent (22). Assembly is slow, and can be assessed by taking points in real time. Disassembly is rapid, and ordinarily requires measurement of tape run at low speed or in a frame-by-frame mode. To avoid operator bias, it is advisable to move the cursor aside between measurements. The frequency with which points are taken depends on the speed of movement of the microtubule tip and on the desired precision. A good rule of thumb is to take points at a frequency that produces an average spacing between them about equal to half the resolution of the microscope or to half the least count of the digitizing board. Sample measurements are shown in *Figure 5*.

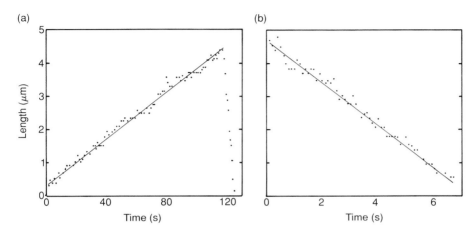

Figure 5. Measurements of the length of a single microtubule, much like that in *Figure 4*, as it undergoes growth and shortening. Measurements over a 120-s period are shown in (a). At approximately 115 s, a shortening process begins. These measurements were taken with the videotape running at real-time speed. Part (b) shows measurements of the shortening process on an expanded time-scale. These measurements were taken on still frames, and every fifth frame was measured.

References

1. Mitchison, T. and Kirschner, M. (1984). *Nature*, **312**, 232.
2. Mitchison, T. and Kirschner, M. (1984). *Nature*, **312**, 237.
3. Horio, T. and Hotani, H. (1986). *Nature*, **321**, 605.
4. Himes, R. H. and Detrich, H. W., III (1989). *Biochemistry*, **28**, 5089.
5. Melki, R., Carlier, M.-F., and Pantaloni, D. (1988). *EMBO J.*, **7**, 2653.
6. Mandelkow, E. M., Lange, G., Jagla, A., Spann, U., and Mandelkow, E. (1988). *EMBO J.*, **7**, 357.
7. Walker, R. A., O'Brien, E. T., Pryer, N. K., Soboeiro, M. F., Voter, W. A., and Erickson, H. P. (1988). *J. Cell Biol.*, **107**, 1437.
8. Caplow, M., Ruhlen, R., Shanks, J., Walker, R. A., and Salmon, E. D. (1989). *Biochemistry*, **28**, 8136.
9. O'Brien, E. T., Salmon, E. D., Walker, R. A., and Erickson, H. P. (1990). *Biochemistry*, **29**, 6648.
10. Cassimeris, L., Pryer, N. K., and Salmon, E. D. (1988). *J. Cell Biol.*, **107**, 2223.
11. Sammak, P. J. and Borisy, G. G. (1988). *Nature*, **332**, 724.
12. Schulze, E. and Kirschner, M. (1988). *Nature*, **334**, 356.

13. Inoué, S. (1981). *J. Cell Biol.*, **89**, 346.
14. Allen, R. D., Allen, N. S., and Travis, J. L. (1981). *Cell Motility*, **1**, 291.
15. Schnapp, B. J. (1986). *Meth. Enzymol.*, **134**, 561.
16. Inoue, S. (1986). *Video Microscopy*. Plenum Press, New York.
17. Ellis, G. W. (1985). *J. Cell Biol.*, **101**, 83a.
18. Inoué, S. (1986). *Video Microscopy*, pp. 127, 496. Plenum Press, New York.
19. Berg, H. C. and Block, S. M. (1984). *J. Gen. Microbiol.*, **130**, 2915.
20. Williams, R. C., Jr. and Lee, J. C. (1982). *Meth. Enzymol.* **85**, part B, 376.
21. Murphy, D. B. (1982). *Meth. Cell Biology*, **24**, 31.
22. Mizushima-Sugano, J., Maeda, T., and Miki-Nomura, T. (1983). *Biochim. Biophys. Acta*, **755**, 257.
23. Williams, R. C., Jr. and Detrich, H. W., III (1979). *Biochemistry*, **18**, 2499.
24. Bell, C. W., Fraser, C., Sale, W. S., Tang, W. Y., and Gibbons, I. R. (1982). *Meth. Cell Biology*, **24**, 373.

8

Methods for the purification and assay of microtubule-associated motility proteins

ROGER D. SLOBODA

1. Introduction

Directed intracellular organelle movement occurs in association with cytoplasmic microtubules in many types of eukaryotic cells, where the microtubules act as a substrate along which the movement occurs (1, 2). Microtubules are long, tubular polymers of indefinite length having an overall diameter of about 25 nm, a hollow core about 15 nm in diameter, and a wall thickness of about 5 nm. They are composed of a single type of subunit protein called tubulin, which exists in its native form as a heterodimer composed of two polypeptides called α- and β-tubulin. Each tubulin subunit is the product of a distinct gene, and all eukaryotes contain at least one gene for each of α- and β-tubulin. The movement of various types of membrane-bound cellular organelles occurs in association with cytoplasmic microtubules. The motive force necessary for such intracellular particle motility is provided by specific proteins that associate with the microtubule subunit lattice in an ATP-dependent manner. The molecules currently known to play a role in microtubule-dependent movements are called kinesin and cytoplasmic dynein.

Section 2 below presents a brief overview of microtubule structure and motility, and Section 3 describes the video microscope technology employed in an analysis of the movements induced by kinesin and cytoplasmic dynein. Section 4 describes how these molecules are purified from cellular homogenates, while Section 5 describes simple *in vitro* assays for microtubule-based motility induced by the purified proteins. The chapter concludes with a section that provides details of several more sophisticated *in vitro* motility assays that reveal not only if a purified fraction has microtubule-dependent motor activity but also the direction in which that motility occurs with respect to the inherent structural polarity of the microtubule.

167

2. Microtubule polarity

A number of years ago Weisenberg (3) reported the conditions in which microtubules could be assembled from cellular homogenates, particularly those derived from nervous tissue such as vertebrate brain. The brain provides a rich source of α- and β-tubulin, because the axons and dendrites of neurons contain many microtubules oriented with their long axes parallel to the long axes of these cytoplasmic extensions. The subunit for assembly of a microtubule is the tubulin dimer, and under appropriate conditions the figure-eight-shaped dimers can be induced to assemble in a head-to-tail fashion producing a three-start, left-handed helix with the long axis of the heterodimer parallel to the long axis of the microtubule (*Figure 1*). This self-assembly reaction generates microtubules that normally contain 13 dimers when viewed in cross-section. Because the α-β heterodimers are asymmetric, the microtubules produced by such asymmetric subunits are asymmetric as well. That is, an assembled microtubule has an inherent structural polarity: one end has the α-subunit of the heterodimer exposed, while the other end has the β-subunit exposed. Microtubules are therefore polar polymers, and thus the addition of subunits during the assembly reaction to each end of such a directional polymer is a different chemical event. In fact, it has been known for a number of years that the two ends of a microtubule are kinetically distinct, because subunits add to them at different rates (4, 5). Under appropriate conditions, subunits add more quickly to one end than to the other. The established convention refers to the end that assembles more

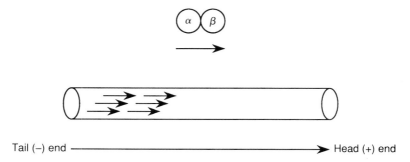

Figure 1. Because the figure-eight-shaped tubulin dimer (top) is composed of two different polypeptide chains, it has an inherent structural asymmetry, as indicated by the arrow beneath the dimer. When these asymmetric subunits polymerize helically in a head-to-tail fashion, they generate a microtubule which is itself asymmetric, thus having a plus (or head) end and a minus (or tail) end. From this diagram it appears that the head end of the microtubule has the ß-tubulin subunit exposed. However, this is an arbitrary designation for diagrammatic purposes only, as it has not yet been determined which subunit is exposed at the head or the tail end of the microtubule.

quickly as the plus or head end, while the end that adds subunits more slowly is referred to as the tail or minus end (*Figure 1*).

The implications of this phenomenon and how it relates to microtubule function in cells have been discussed by Kirschner (6) and by Mitchison and Kirschner (7); interested readers are referred to the large and growing literature on what has been termed dynamic instability by Mitchison and Kirschner (7) to describe this phenomenon. For the purposes of understanding the assays that have been developed for studying microtubule-based motility motors, one need only realize that, because a microtubule has a head and a tail, translocation of a particle from the minus to the plus end is a different molecular event than movement of a particle from the plus to the minus end, due to the asymmetric nature of the underlying tubulin dimers and hence the subunit lattice produced upon their polymerization.

The determination of the structural polarity of microtubules based on kinetic results as mentioned above immediately caused investigators to question the manner in which microtubules are organized within cells. Because most, if not all, microtubules originate from an organizing centre—a centriole with a surrounding cloud of pericentriolar material—experiments were designed to determine the polarity of microtubules with respect to this organizing centre. That is, are all microtubules arranged parallel (e.g. with the same end distal to the organizing centre), anti-parallel, or random? By using various solution conditions in conjunction with specific proteins as morphological probes, several investigators established methods for determining microtubule polarity. These procedures involve the incubation of fixed and permeabilized cells with solutions containing tubulin subunits (8) or flagellar dynein (9). Under the conditions employed, the tubulin or dynein subunits bind to the walls of existing microtubules, decorating them in a pinwheel fashion as observed when the microtubules are subsequently viewed in cross-section. The direction the arms of the pinwheels face (clockwise or counterclockwise) defines the underlying structural polarity of the tubulin subunit lattice. When applied to interphase tissue culture cells, it was noted that all microtubules emanate from the centrosome with their plus ends distal and their minus ends proximal to the organizing centre; hence they are parallel, with their fast-growing ends facing the cytoplasm. When similar approaches were applied to the axons of neuronal cells, it was determined that all the microtubules were oriented with their plus ends distal to the cell body (10), i.e., facing the synapse. During mitosis, the minus ends of the microtubules point toward the pole of each half spindle, while the plus ends of microtubules from a given half spindle are either attached to the kinetochores of the chromosomes or interdigitate at the centre of the spindle with plus ends of microtubules projecting from the cognate half spindle (9, 11). Thus, in the mitotic apparatus each respective half spindle contains parallel microtubules, while the spindle as a whole also contains anti-parallel microtubules at its centre, where the microtubules emanating from each pole overlap.

169

Thus, the determination of the orientation of cytoplasmic microtubules with respect to their structural polarity has formed the basis for the design of many experiments on microtubule motors. Directed intracellular particle movement therefore depends on different motor molecules that interact with the microtubule subunit lattice in such a manner that movement is effected in only one direction, by convention termed plus-end-directed (kinesin-driven) or minus-end-directed (cytoplasmic dynein-driven).

3. Particle motility and video microscopy of microtubule-dependent movement

3.1 Microtubule-dependent intracellular particle motility

It has been known since the early work of Weiss and Hiscoe (12) that transport of material occurs in axons; these authors named this process axonal flow. The initial observations have been expanded and refined in the ensuing years by the application of isotopically labelled precursors to the study of axonal flow. In these studies (see 1, 13 for reviews) radioactive amino acids were injected into cell bodies or ganglia, and the nerves were then excised, sectioned, and analysed at various times after the injection. Initial experiments demonstrated a wave of radioactivity moving down the axon as a function of time after injection. Similar subsequent experiments that coupled this approach to analysis by gel electrophoresis and autoradiography allowed the protein components to be determined along with their relative rates of movement. Thus, microtubule-dependent transport in axons can be classified into three major groups, termed fast, intermediate, and slow transport. Each group is further subdivided on the basis of the biochemical and morphological characteristics of the components being moved, their relative rates of movement, and their direction with respect to axonal morphology (13). This latter characteristic—directionality of movement—is the one most relevant to current studies of the methods for the assay of microtubule-based motors under discussion here.

Fast axonal transport refers to the bidirectional movement of membrane-bound organelles in association with microtubules. This movement occurs from the cell body to the nerve terminal (the orthograde or anterograde direction) and from the nerve terminal to the cell body (the retrograde direction) along microtubules in the axon. Newly synthesized membranous organelles produced in the cell body are transported down axonal micro-tubules in the orthograde direction toward the synapse. The organelles that are moving in the retrograde direction, however, are different in nature and are being transported from the synapse back to the cell body for degradation or recycling (reviewed in 1, 13–15). Recall, however, that this bidirectional transport system occurs on a microtubule substrate that is unidirectional; that is, all of the microtubules have their plus ends pointing toward the nerve terminal. The manner in which bidirectional transport on a unidirectional substrate is accomplished was the subject of intense research activity during the 1980s, and it was discovered several years ago that the motive-force-generating proteins kinesin (16) and cytoplasmic dynein (17) play major roles in this process. The directed transport of intracellular membrane-bound organelles in almost all cells is now thought to be accomplished by members of these two families of proteins. Identification, isolation, and purification of these proteins was dependent on the concurrent development of new methods of computer-enhanced microscopy, because this instrumentation enabled an *in vitro* assay to be developed for the microtubule-dependent motor activity of these proteins. A knowledge of how the modern video microscope functions and how its important components are assembled is essential to understanding and performing the assays for microtubule-dependent motors described in the following sections.

3.2 Video microscopy

A major advance in the study of microtubule-based intracellular particle transport occurred in the early 1980s with the application of video and digital image processing techniques to standard differential interference contrast (DIC) microscopes. Until this time state-of-the-art DIC microscopes using conventional 16 mm movie or 35 mm still cameras were not capable of imaging microtubules and the movements of particles along them, being constrained by Abbé's relationship and the laws of physics to a theoretical limit of resolution of around 200 nm. This is almost an order of magnitude greater than the diameter of a microtubule (25 nm), yet these organelles are rendered visible (note that they are not resolved, however, by this technique) because of several unique features of the system that greatly increase visual contrast.

DIC microscopy (18, 19) produces shadowcast images because the microscope generates contrast due to variations in refractive index gradients. For example, at the edges of cells or at the edges of cellular organelles bound by internal membranes the rate of change of refractive index is steepest and thus the most contrast is generated, as evidenced by the steepness of the resulting 'shadows' observed in the images. To generate optimal contrast, the compensator is usually set to within $\pm$ 0.1 λ of extinction, and the iris diaphragm in the condenser is stopped down, thus sacrificing optimal resolution for optimal contrast. In video DIC microscopy (20, 21), however, the iris diaphragm is opened to match the numerical aperture of the objective, and the compensator is set in the range of $\frac{1}{2}$ λ–$\frac{1}{4}$ λ of extinction. Under these conditions, the microscope image contains so much stray light that the specimen may not be visible to the naked eye or to normal photographic film. An appropriate video camera can deal with such images, however, because of the camera's ability to reject stray light, thus greatly increasing contrast. This is accomplished by gain and offset controls which are used (see steps below) in effect to clamp the voltage in a DC restoration circuit. For example, in normal DIC microscopy, the contrast (C) is given by the following formula (20):

$$C = (I_b - I_s)/I_b \qquad (1)$$

where I_b is the intensity (or brightness) of the background and I_s is the intensity of the specimen. The video DIC microscope can accomplish a reduction in I_b without reducing $I_b - I_s$. The result is increased contrast (20, 21):

$$C_{video} = (I_b - I_s)/(I_b - I_v) \qquad (2)$$

where I_v is the reduction in signal due to the variation of the clamp level of a DC restoration circuit in the video camera (20, 21). In effect, I_v subtracts from the video signal a voltage that corresponds to the excess background light that is the cause of the low contrast problem in the first place. The knob controlling this voltage may be called the offset control, the black level, or the threshold, depending on the manufacturer of the camera. The contrast (C) is thus greatly increased, according to Equation 2, in which I_v (in effect, a negative intensity) produces a decrease in the ($I_b - I_v$) term of the equation, thus resulting in increased C. Once the background is appropriately reduced, the signal is then increased by increasing the gain control of the video circuit.

A second series of steps using digital image-processing techniques (22) is then required to generate an optimal image. The use of such high amplification procedures as outlined above enhances not only image contrast but also lens imperfections and particles of dust or dirt that may exist on each of the prism, mirror, and lens surfaces in the optical path. These are all projected on the television image, producing a pattern called 'mottle' by Allen and Allen (23). Thus, the next step in video DIC microscopy is to send the video image to a computer frame memory. To do this, the specimen is brought out of focus; the mottle remains visible and unchanged and is continually sent to the camera even when the specimen is not in focus. The image of the mottle is stored in the computer, and the computer is then instructed to subtract the background mottle in real time from each succeeding in-focus image sent to the computer by the video camera. The background-subtracted image is then further contrast-enhanced by digital image-processing techniques, which in effect expand the relative gray scale of the image being sent to the computer by the camera. The final result is a clearly defined image having adequate contrast. In addition, these techniques are particularly suited to the study of directed intracellular particle transport in general and to microtubule-based motility in particular because of the frame averaging capability of the computer. In this mode, the system is set up as outlined above, and then several video frames can be averaged prior to their display on the video monitor. This process tends to remove from the video image random movements, which are decreased or lost completely due to the frame averaging mode, while directed movements (i.e., particle translocations) are correspondingly enhanced. For a detailed overview of video microscopy techniques, interested readers are referred to the excellent text by Inoué (24).

3.3 Setting up the video DIC microscope

The microscope should be equipped with a 100W mercury (Hg) arc lamp for illumination, a heat-absorbing filter and a 546 nm interference filter (to render the illumination monochromatic), a high-resolution condenser and objective (each with a numerical aperture of 1.3 or greater), a quarter-wave plate, and a rotatable analyser. The specific steps in producing an appropriate image will depend on the type of DIC microscope, video camera, and computer system being used (see following section). In generic terms, however, the steps are performed in the sequence shown in *Protocol 1*, adapted almost directly from Allen and Allen (23).

Protocol 1. Setting up the video DIC microscope

1. Focus the specimen for optimal viewing by eye, using the principles of Köhler illumination (see for example Spencer ref 25). Note, however, that most microscopes available off the shelf are not capable of filling the front focal plane of the condenser with an image of the mercury lamp arc, one of the steps in obtaining Köhler illumination. Thus, in practice, for video DIC microscopy the microscope is operated under conditions of critical illumination, in which case the image of the mercury arc is focused in the specimen plane, not the front focal plane of the condenser. This step is required to provide the high level of illumination necessary for video DIC microscopy, which must be capable of almost saturating the tube of the video camera.

2. Rotate the analyser (which should be preceded in the optical path by a $\frac{1}{4} \lambda$ plate) up to 20–45 ° from extinction. This value will depend on overall image intensity (see step 3 below). Open the iris diaphragm in the condenser to exceed slightly the numerical aperture of the objective. These two steps should severely degrade the visual image due to the great increase in background light and the accompanying decrease in contrast (Equation 1).

3. The microscope is then manipulated to send the image to the video camera. Be careful not to send so much light to the camera that it becomes saturated (usually indicated by a warning bell, light, etc.). If this occurs, the analyser has probably been rotated too far from extinction, in

Protocol 1. *Continued*

which case it should be brought back towards extinction until the camera is no longer saturated.

4. While viewing the image on the monitor, manipulate the offset and gain controls until visual contrast has been optimized.

5. Defocus the image, and store the image of the mottle pattern that remains in the computer according to the manufacturer's instructions. Then instruct the computer to begin subtracting the image of the mottle from each succeeding frame before it is sent to the monitor. With the microscope still out of focus, the monitor should then go blank (i.e., grey). Bring the specimen back into focus.

6. Add electronic contrast (expand the image grey scale according to the manufacturer's instruction) as required. Store the data by recording on 19 mm (¾ inch) or super VHS videotape (cheap) or on optical memory disc (very expensive).

7. In the case of *in vitro* assays for motor activity, frame averaging is usually not required, although time-lapse recording may be helpful in the case of motor molecules that move things slowly. When viewing motility in intact cells or nerve axoplasm or when using certain *in vitro* assays (see for example Sections 5.1, 5.2 or 6.1), however, the frame-averaging capability of the computer is helpful because of its ability to average out and thus greatly reduce random movements of particles caused by Brownian motion.

3.4 Components of a typical video DIC microscope

For a number of years, my laboratory has employed the following components for video DIC microscopy (see for example refs. 26, 27). The optical system includes an upright (not inverted) Nikon Optiphot microscope with a 100W Hg arc lamp for illumination, heat and green interference filtres, an achromatic/aplanatic DIC condenser having a numerical aperture (NA) of 1.35, and a 100 × 1.35 NA planapochromatic objective. Above the objective a DIC Wollaston slider is fitted, and above that a quarter-wave plate (for 546 nm light) is set. The final necessary component is a rotatable analyser which is used to set the microscope at or away from extinction (see Step 2, *Protocol 1* above). The microscope is fitted with a transfer head that allows 100 per cent of the light to be sent to the video camera. The image is transferred from the microscope by a 10–20 × Swift zoom eyepiece set at approximately 16× and focused onto the image plane of a Hamamatsu C-1000 video camera having a Chalnicon video tube using a Zeiss adaptor fitted to the faceplate of the video camera. The adaptor contains a projection lens having a focal length of 63 mm with a T2 thread plus a T2 to C-mount adaptor. The video signal from the camera is then sent to a Photonics C1966 image processor, and enhanced images are recorded in real time using a Sony VO-2610 19 mm (¾ inch) videocassette recorder.

This overview has, of necessity, been succinct. Readers who require more information are referred to the specific review by Schnapp (28) in which detailed procedures for viewing single microtubules by video DIC microscopy are provided.

3.5 Alternatives

The number of necessary components detailed above immediately indicates that video microscopy is expensive, and also that numerous combinations of equipment from various manufacturers can be used (see also Chapter 7). For example, we also use a Zeiss IM35 inverted microscope with much success. In that configuration a 100W Hg arc provides the illumination, and a fixed (i.e., a non-turret condenser with a single position) 1.4 NA DIC condenser and a 63 × 1.4 NA objective generate the image. The image is sent from the microscope to the camera via a 16× Zeiss projection eyepiece. The image leaves the projection eyepiece and travels through a simple cardboard tube (used to block background room illumination from entering the optical path) directly to the camera without the use of an intervening camera adaptor.

For each of the two microscopes described above, the position of the camera was adjusted until the magnification of the resulting image on the video monitor was about 9500×. This translates into a field width of about 18.5 μm on a high-resolution monitor. The new line of Zeiss microscopes (Axiophot, Axiovert) also produce fine video DIC images. Of particular note with this line of instruments is the fact that we and others have successfully viewed microtubules and performed the motor assays described in following sections using the Zeiss Axiovert with simple tungsten or halogen illumination.

Many high-quality video cameras are available on the market, as are a number of digital image-processing devices. They are too numerous to detail here, but a few general principles should guide their purchase. First, the camera should provide high resolution, both spatial as well as temporal, and be capable of providing the necessary DC clamp voltage circuitry to increase contrast (see Equation 2) via controls labelled offset, black, or threshold, etc., depending on the manufacturer. Second, the image-processing system should be capable of background subtraction as well as frame averaging capabilities. That is, it should have more than a single frame memory to enable it to carry out background subtraction and frame averaging simultaneously.

4. Biochemical details of the purification of microtubule-based motor proteins

Two classes of force-generating motor proteins are now recognized as components of numerous different cell types. These proteins are kinesin, a plus-end-directed motor, and cytoplasmic dynein, a minus-end-directed motor. Both proteins generate force using Mg-ATP as an energy source. This section begins with a brief overview of the strategies involved in purifying these proteins, and then provides an outline of the isolation procedures employed; Sections 5 and 6 explain the methods used to test for microtubule-dependent motor activity *in vitro*.

The major approach used in the procedures detailed here has come to be called a fishing experiment using taxol-stabilized microtubules. Taxol is a drug isolated from the western yew, *Taxus brevifolia*, which has the unique property that it drives the microtubule assembly reaction toward the production of polymer by stabilizing assembled microtubules (29). Taxol is currently available only gratis by writing to the Natural Products Branch of the National Cancer Institute and requesting a 20–50 mg sample of the dry powder. The dry form should be specified, as otherwise the taxol may be shipped in an oil-based solution which is of no use for the experiments described here.

For a fishing experiment, microtubules stabilized by taxol are added to a cellular homogenate (or any mixture of proteins) to cause specific microtubule-binding proteins to associate with the microtubules. The microtubules can then be sedimented out of solution with the associated microtubule-binding proteins (the fishing step). Because kinesin and cytoplasmic dynein bind to microtubules in an ATP-sensitive manner, the methods also include an ATP-dependent release step to purify the motor protein away from the taxol-stabilized microtubules after the initial fishing step. Thus, for most of the methods mentioned in this and the following sections, a source of purified tubulin dimers used to form taxol-stabilized *in vitro*-assembled microtubules is essential. Tubulin is normally purified by phosphocellulose chromatography (30) or DEAE chromatography (31). Numerous methods are available in the literature describing these procedures. For details the reader is referred to Chapter 7 or to *Methods in Enzymology*, Volumes 85 and 134.

4.1 Kinesin

The key to the isolation of kinesin from tissue homogenates lies in the fact that the protein binds extremely tightly to microtubules in the presence of the non-hydrolysable ATP analogue 5′-adenylyl imidodiphosphate (AMP-PNP) (32). The procedural details for the isolation of kinesin presented here have been compiled from a number of original papers in the literature (16, 33, 34). Basically, the tissue is homogenized in an aqueous buffer at neutral pH. Because several of the steps occur at temperatures above 4 °C, one or more protease inhibitors are included to prevent polypeptide hydrolysis by endogenous proteases that may exist in the homogenate. For the steps described in this and following sections, all temperatures are 4 °C unless otherwise instructed.

Protocol 2. Isolation of kinesin

1. Homogenize the tissue at a ratio of 1 gm/1.5 ml PMEG buffer: 100 mM Pipes, pH 6.8, 1 mM EGTA, 2 mM $MgSO_4$, 0.1 mM GTP, 1 µg/ml leupeptin, and 1 mM phenylmethylsulphonyl fluoride (PMSF).

2. Centrifuge the homogenate at 39 000 × g for 30 min. Remove the supernatant and discard the pellet.

3. Centrifuge the supernatant from step 2 at 100 000 × g for 1 h.

4. Adjust the supernatant from step 3 to 20 µM taxol and incubate at 30 °C to promote microtubule assembly.

5. Collect the microtubules and endogenous microtubule-binding proteins (MAPs) by centrifugation at 39 000 × g for 30 min at 4 °C. Because kinesin is more readily dissociated from microtubules in the presence of the nucleotides ATP or GTP, the bulk of the kinesin remains soluble at this step.

6. Add apyrase (EC 3.6.2.5) to the MAP- and tubulin-depleted supernatant from step 4 to a final concentration of 2 U/ml and incubate at 25 °C for 30 min. This step depletes the solution of endogenous ATP, which is hydrolysed by the apyrase to ADP and then AMP.

7. While the above incubation is proceeding, polymerize an appropriate amount (see step 8) of previously purified tubulin dimers (obtained as described in Chapter 7) by incubation at 25 °C in PMEG buffer containing 0.1 mM GTP and 20 µM taxol for 20 min.

8. Adjust the apyrase-treated solution from step 6 to a final concentration of 0.1 mg/ml microtubules (from step 7) and 0.5 mM AMP–PNP. Incubate at 25 °C for 30 min. The kinesin binds tightly to assembled microtubules in the presence of AMP–PNP.

9. Collect the microtubules and kinesin by centrifugation at 39 000 × g for 30 min, discard the supernatant, and resuspend the pellet in PMEG containing 0.1 mM AMP–PNP and 20 µM taxol. This step 'washes' the microtubules and removes any proteins trapped non-specifically in the microtubule pellet.

10. Collect the microtubules again by centrifugation at 39 000 × g for 30 min.

11. Resuspend the pellet of microtubules from step 10 in PMEG containing 20 µM taxol and 10 mM Mg-ATP, and incubate at 25 °C for 30 min with occasional agitation brought about by gentle inversion or drawing the solution up and down with a Pasteur pipette. This incubation (the ATP release step) dissociates the kinesin from the microtubules, which can then be separated from the solubilized kinesin by centrifugation as above.

Protocol 2. *Continued*

The resulting supernatant is greatly enriched in kinesin. The kinesin at this step is adequate for use in the motor assays that are described in the following sections. If required, kinesin can be purified further by ion exchange chromatography on DEAE or phosphocellulose, by molecular sieve chromatography using either Bio Gel A-5m or Sephacryl S-300, or by a combination of both approaches. The details of these further purification procedures can be found in Kuznetsov and Gelfand (33), Bloom *et al.* (34), and Hackney (35).

4.2 Cytoplasmic dynein

The purification of cytoplasmic dynein follows the same strategy as outlined above for kinesin, employing a microtubule affinity step followed by an ATP release step. Buffer components are the same, and only slight variations occur to enrich for dynein while at the same time excluding kinesin from the final preparation. The steps outlined in *Protocol 3* below are condensed from a compilation of approaches previously used by others (36, 37), and are based on dynein purification strategies originally developed for axonemal dyneins.

Protocol 3. Isolation of cytoplasmic dynein

1. Homogenize the tissue in PMEG containing 1 mM GTP at a ratio of 1 gm tissue/1.5 ml PMEG. Spin the homogenate at 39 000 × *g* for 30 min. Discard the pellet.

2. Centrifuge the supernatant from step 1 at 100 000 × *g* for 1 h. Discard the pellet.

3. Add apyrase to the supernatant from step 2 to a final concentration of 2 U/ml and incubate at 25 °C for 30 min to deplete the ATP.

4. While the above incubation is proceeding, polymerize some previously purified tubulin dimers (obtained as described in Chapter 7) by incubation for 20 min at 25 °C in PMEG buffer containing 1 mM GTP and 20 μM taxol.

5. Adjust the ATP-depleted solution from step 3 to 0.1 mg/ml microtubules and 20 μM taxol, and incubate for another 30 min at 25 °C. During this incubation the cytoplasmic dynein binds to the taxol-stabilized micro-tubules; the kinesin does not, because it is prevented from binding by the presence of the 1 mM GTP added to the original homogenate. (Note that when working with a tissue that is rich in tubulin such as neuronal cells, the endogenous tubulin concentration will be sufficiently high that there is no need to add extra purified dimers at this step. In this case simply polymerize the endogenous tubulin with taxol and GTP.)

6. Collect the microtubules and bound cytoplasmic dynein by centrifugation at 39 000 × *g* for 30 min, discard the supernatant, and resuspend the pellet in PMEG containing 1 mM GTP and 20 μM taxol. This step 'washes' the microtubules and removes any proteins trapped non-specifically in the microtubule pellet.

7. Collect the microtubules again by centrifugation as before.

8. Resuspend the pellet of microtubules from step 7 in PMEG containing 20 μM taxol and 10 mM Mg-ATP, and incubate at 25 °C for 30 min with occasional agitation brought about by gentle inversion or by drawing the solution up and down with a Pasteur pipette. This step dissociates the cytoplasmic dynein from the microtubules; the microtubules can then be removed from the solubilized cytoplasmic dynein by centrifugation as above.

9. Finally, layer the ATP-released supernatant containing cytoplasmic dynein on a 5–20 per cent sucrose gradient in PMEG and centrifuge at 175 000 × *g* for 16 h in a Beckman SW41 rotor. The cytoplasmic dynein will elute from the gradient at a position equivalent to 19–20S.

4.3 General comments

It is readily apparent from the above procedures that both approaches can be combined in a single protocol yielding kinesin and cytoplasmic dynein from the same starting homogenate. However, in daily practice this is normally not done, as a given laboratory is usually studying either one protein or the other. However, a combined protocol has been used successfully by the author for instructional purposes with graduate and undergraduate students, and such a procedure is outlined in the flow chart shown in *Figure 2*.

5. Determination of microtubule-based motor protein activity

The assays in this section as well as in Section 6 are based on the translocation of marker particles along the surface of microtubules (vesicle or bead movements) or on the translocation of microtubules along a glass surface (microtubule gliding). Both of these approaches depend on high-resolution video DIC microscopy to visualize the microtubules and the motility associated with them in real time. In effect, both assays are dependent on the movement of an object relative to a microtubule; in the former case, the object containing the motor protein on its surface translocates on a stationary microtubule, while in the latter case, the microtubule glides on a stationary substrate (the glass slide) coated with the motor protein (*Figure 3*). This

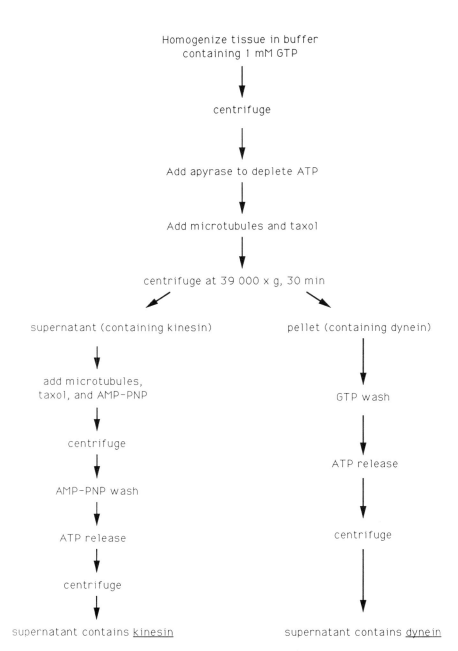

Figure 2. Flow chart for the isolation of kinesin and cytoplasmic dynein from a single starting homogenate. The initial separation depends on the differential binding of kinesin and cytoplasmic dynein to microtubules in the presence of GTP. Note that the combined procedure outlined here produces fractions that are greatly enriched for kinesin or cytoplasmic dynein, but that further purification steps would be required to purify the proteins to homogeneity. See text, Sections 4.1 and 4.2, for complete details.

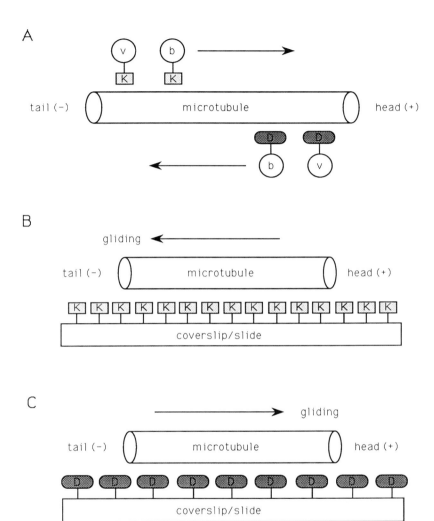

Figure 3. Characteristics of the motor assays described in the text. *A*: Membrane-bound vesicles (v) or latex beads (b) with kinesin (K) attached will translocate toward the head (or plus) end of the microtubule, while particles with dyenin (D) attached will move toward the tail (or minus) end of the microtubule. *B*: A microtubule will glide along a glass surface containing bound kinesin (K) with its plus (or head) end trailing. This is functionally equivalent to the movement of the slide toward the plus end of the microtubule. *C*: A microtubule will glide along a glass surface containing bound dynein (D) with its minus (or tail) end trailing. This is functionally equivalent to movement of the slide toward the minus end of the microtubule.

section will provide the methods for assaying the presence of motor activity in general, while the following section (Section 6) will describe several ingenious techniques that have been developed to assess motility with respect to microtubule polarity (i.e., plus end- vs. minus end-directed motors).

5.1 Movements of intact membrane vesicles

This approach, based on the work of Gilbert *et al.* (26), was first developed to determine whether membrane vesicles from squid axoplasm carried their own motor proteins or whether the vesicles moved by interacting with motor proteins that resided permanently on the surface of the microtubules. A simple analogy will suffice: do the particles move on microtubules like a locomotive on a railroad track or like people on an escalator? In the former case, the motor is associated with the moving particle (the locomotive), while in the latter case the motor is a component of the track (the escalator). The assay is simple in its approach, but can be terribly complex in its execution, simply because of the large number of variables possible in the formulation of buffer components, etc. The assay that works well for squid axoplasm is outlined below and is based on Gilbert *et al.* (26). As before, all steps in this and following sections are at 4 °C unless instructed otherwise.

Protocol 4. Assay for movement of squid axoplasmic vesicles

1. Homogenize the tissue in SKEM buffer (250 mM sucrose, 250 mM KCl, 10 mM $MgCl_2$, 1 mM EGTA, 1 mM ATP, 1 mM PMSF, 10 mM imidazole, pH 7.4) at a ratio of 1 ml SKEM per gram tissue, and centrifuge the homogenate at 12 000 × *g* for 30 min.

2. Layer the supernatant over a discontinuous, three-step sucrose gradient containing 0.3, 0.6, and 1.5 M sucrose in SKEM (lacking the sucrose and with the KCl reduced to 100 mM), and centrifuge for 5 h at 58 000 × *g* in a Beckman SW50.1 rotor. Vesicles capable of motility in association with microtubules band at the 0.3–0.6 M sucrose interface.

3. Isolate substrate microtubules for this assay from sea urchin sperm flagella using the procedure described in Chapter 7, *Protocol 2*.

4. For *in vitro* motility, place an aliquot of washed axonemes in Hepes buffer on an 18 × 18 mm #0 cover slip and allow them to adhere by settling for 2–3 min. Remove unattached axonemes by washing the cover slip with a few drops of Hepes buffer, and insert the cover slip on to a 24 × 60 mm #0 cover slip. Seal the cover slip sandwich on the front and back edges with valap, attach it to a metal holder specially designed for this purpose (39), and transfer it to the video DIC microscope for observation.

5. Add a few μl of vesicles from step 3 above to one side of the cover slip

sandwich and wash them across the field of outer doublets by absorbing buffer with filter paper at the opposite side.

6. Observe the preparation with the video DIC microscope, and note the directed movements of vesicles along the outer doublet microtubules.

This type of assay has the advantage that the presence of motor proteins in association with membrane-bound vesicles can be, theoretically at least, readily assessed. In the case of vesicles from squid axoplasm, it works routinely. However, one should note that squid axoplasm is far from a routine preparation. Material is obtained by dissection of the axon and then extrusion of the axoplasm from the surrounding membrane and glial sheath. Thus, all of the vesicles isolated by the sucrose step gradient presumably comprise representatives of the entire vesicle population of the axon, the vast majority of which were in active movement at the time of the extrusion. This is not the case with, for example, a tissue homogenate in which the bulk of the membrane vesicles obtained result from vesiculation of plasma membrane and endoplasmic reticulum caused by the homogenization. Thus, when applied to other cell types, these procedures would presumably yield a preparation of vesicles in which only a small percentage of vesicles in the microscope field were actually carrying motor proteins on their surface and were capable of moving on a microtubule substrate. Moreover, the required buffer conditions for other systems can be quite different in parameters such as pH, ionic strength, ionic cofactors, etc. Thus, much experimentation may be required for other systems to obtain a vesicle preparation that retains motor activity. Hence, other approaches have also been developed to circumvent these problems. These approaches involve enriching or purifying the putative microtubule motor proteins using the procedures outlined for kinesin and cytoplasmic dynein in Section 4, and then testing their ability to promote motility using one of the following assays.

5.2 Movements of latex beads coated with motor proteins

This assay, which is modified for microtubules from an approach first utilized by Sheetz and Spudich (40) for studying myosin-based movements on actin cables from the green alga *Nitella*, involves the coating of polystyrene or latex microspheres with a solution containing presumptive motor activity. The motor protein adsorbs to the bead surface, producing in effect an artificial vesicle analogous to the membrane vesicle/motor protein particles described in the preceding Section 5.1 (see also *Figure 3*). The remarkable thing about this assay is that, although the motor proteins presumably bind to the bead surface randomly, enough proteins do so with their microtubule binding domains oriented appropriately that the motor protein can effect movement of the beads along a microtubule substrate. The steps employed in the assay, adapted from Vale *et al.* (41), are outlined in *Protocol 5*.

Protocol 5. Assay for movement of motor-protein-coated latex beads

1. Incubate all cover slips (#0's, both top and bottom used to form the sandwich similar to that described in *Protocol 4*) to be used in this assay overnight in a solution of 200 µg/ml poly-D-lysine and wash each cover slip four times with 50 ml distilled water. Air dry.

2. Dilute carboxylated latex beads (purchased as a 2.5 per cent suspension) 300-fold into a suitable motility buffer. Originally, the composition of this buffer (16) was half the strength of buffer X, which was developed by Lasek and Brady (32) to mimic the composition of squid axoplasm; however, with experience it has been determined that the composition of the buffer is not so critical, and a less complex buffer is now routinely employed. For example, the PMEG buffer described in Section 4.1, supplemented with 1 mM ATP and 1 mM DTT, supports motility quite readily. For completeness, however, the motility buffer originally employed in the identification of kinesin from squid axoplasm was 175 mM potassium aspartate, 65 mM taurine, 85 mM betaine, 25 mM glycine, 20 mM Hepes, pH 7.2, 6.4 mM MgCl$_2$, and 5 mM EGTA, containing in addition 20 µM taxol, 1 mM GTP, and 2 mM ATP (16). Excluding the additions, this is essentially half-strength buffer X, with the exception that Vale *et al.* (16) used 85 mM betaine, while half-strength buffer X would contain 35 mM betaine and additionally 1.5 mM CaCl$_2$, 0.5 mM glucose, and 0.5 mM ATP, according to the formulation employed by Lasek and Brady (32).

3. Mix the diluted beads with the solution containing the presumptive motor protein (in the same buffer) at a ratio of 1 part beads to 4 parts protein, and incubate for 5 min on ice.

4. Meanwhile, induce purified tubulin dimers at 0.5 mg/ml to assemble by the addition of taxol and GTP, as described in Section 4.1.

5. Mix 1 part microtubules with 3 parts motor-coated latex beads, apply 5 µl of this solution between two #0 cover slips as described in the preceding section, and seal with valap.

6. Observe the preparation with video DIC to visualize translocations of the beads along the microtubules. To do this, focus on microtubules that are lying attached to and thus immobilized on the cover slip. Unbound beads will be observed floating in solution above the microtubules; thus, it is also helpful to use the frame-averaging capability of the computer to average out this random motion, making directed translocations along the microtubules much more obvious.

5.3 Microtubule gliding

By far, the most straightforward and easily performed motor protein assay is microtubule gliding, in which taxol-stabilized microtubules glide along the glass substrate due to the action of motor proteins bound to the glass surface. The assay is analogous to a motor-coated bead moving along a microtubule (Section 5.2), in that the cover slip can be thought of as a single huge glass bead. Instead of translocating along the microtubule, however, the microtubule translocates along the immobile glass surface (see *Figure 3*). The assay works well with both purified kinesin and purified cytoplasmic dynein, and has even been used to assay for the presence of motor activities in crude homogenates. Indeed, the gliding assay was employed in the initial biochemical fractionation experiments that led to the discovery of kinesin (16), and is based on an observation made initially with extruded squid axoplasm. It was first noted, when extruded axoplasm was observed using video DIC, that microtubules would move along the glass cover slip (42) or on a cover slip coated with a supernatant obtained from an axoplasmic homogenate (41). Using this system as an assay, these latter authors first identified kinesin. In practice, the gliding assay is quite simple in execution, yet it is extremely elegant and powerful in its ability to reveal the presence and identity of a microtubule-based motor.

Protocol 6. Assay for microtubule gliding

1. Construct a humid chamber by lining a Petri dish with a layer of moist filter paper. Place an 18 × 18 mm #0 cover slip in the chamber. The cover slip should *not* have been treated with poly-D-lysine. (Be sure the filter paper is not so moist that liquid runs up on to the surface of the cover slips.)

2. Add to the cover slip 50 μl of a solution containing either purified kinesin, purified cytoplasmic dynein, or the presumptive microtubule motor protein in PMEG buffer. Place the lid on the Petri dish, and allow the protein to adhere to the cover slip for about 5 min at room temperature.

3. Rinse the cover slip briefly with a few drops of fresh PMEG buffer, invert it on to a 24 × 60 mm #0 cover slip, and then seal the front and back with valap as described above.

4. To one side of the cover slip sandwich, add 50 μl of taxol-stabilized microtubules at 0.5 mg/ml in PMEG buffer containing 2 mM Mg-ATP, and perfuse the microtubules between the cover slips by drawing buffer out the other side using capillary action and a piece of filter paper. Allow this to sit for 3–5 min in the humid chamber to give the microtubules a chance to adsorb to the cover slip, and then repeat with a fresh 50 μl aliquot of microtubules.

185

Protocol 6. *Continued*

5. Wash through the cover slip sandwich about 200 μl fresh PMEG buffer containing 2 mM ATP to remove any non-bound microtubules, mount on the microscope stage, and observe with video DIC.

Alternatively, another approach to this technique is to add sequentially to a single cover slip the various components, i.e., first add the solution of motor protein, remove by washing, add the microtubules, remove by washing, and then mount on a second cover slip, seal with valap, and observe. Kinesin-driven microtubule gliding will occur at a rate of 0.3–0.5 μm/s (16), while motility due to cytoplasmic dynein will be faster, at rates approaching 1.25 μm/s (36). By contrast, gliding rates due to axonemal dyneins can be 4–8 μm/s (43). Consult also McIntosh and Porter (44) for a recent compilation of studies on the biochemical and motility characteristics of various known microtubule-based motors. Note that these rates indicate, in the case of kinesin for example, that a gliding microtubule could take around 1 min to traverse a single microscope field (recall that the video DIC microscope as routinely operated (see Section 3.5) at a total magnification of 9500× generates a field width of 18.5 μm on the video monitor). Thus, the user should be aware that gliding, while usually consistent, may not be immediately apparent until one 'learns' to recognize the characteristics of the motility. Initially, one might find it helpful to record several fields first in time-lapse mode and then in real time. In this manner the time-lapse sequences can be played back at increased rate to make movements readily obvious. It therefore becomes easy to determine which microtubules are moving. Then the same microtubules can be examined in real time to get the observer's eye accustomed to the characteristics of the movements. If time-lapse equipment is not available, it is possible to watch a single microtubule on the monitor, marking its position with a felt-tipped pen if necessary. After a while the movements become readily discernible.

6. Specific motility assays that reveal directionality with respect to the inherent polarity of microtubules

Ingenious methods have been devised that modify several of the previously outlined procedures to enable one to determine the directionality of movement with respect to the head-to-tail polarity of microtubules. These involve the movements of latex beads on polarized microtubule substrates of known directionality, the gliding of microtubules that have had extra subunits added to their plus ends to reveal their polarity, and the gliding of sperm tails of known polarity. Each of these procedures will be discussed in turn below.

6.1 Movement of plastic beads on microtubules nucleated from isolated centrosomes

In their experiments on the dynamic instability of microtubules, Mitchison and Kirschner (7) developed a method for nucleated growth of purified tubulin dimers off of centrosomes isolated from tissue culture cells (45). Microtubules in cells grow from centrosomes with the plus ends of the microtubules distal to the centrosome, as reviewed in Section 1. Based on the kinetics of microtubule elongation from isolated centrosomes it has been determined that the *in vitro* situation is identical: centrosomes nucleate microtubules with the plus ends of the microtubules distal to the centrosome. Thus, the centrosome and associated microtubules provide an oriented substrate for *in vitro* motility assays in which the polarity of the microtubules is known. It is a relatively straightforward task to determine the directionality of movement of a given motor protein using this assay. What is not so straightfoward is the expertise and added work required to purify centrosomes, polymerize oriented microtubule arrays from them, and then attach these complexes to cover slips for viewing, etc. For this reason other assays have also been developed that can be performed more readily. In addition to the centrosome assay discussed here, three other approaches having varying degrees of complexity are also provided in Sections 6.2–6.4 below.

The first approach to be outlined here combines the latex bead assay (see Section 4.2) with a polarized microtubule substrate nucleated from purified centrosomes. The following steps have been adapted from the work of Mitchison and Kirschner (45) and of Vale *et al.* (46).

Protocol 7. Assay for movement on centrosome-nucleated microtubules

The isolation procedure for centrosomes is described in clear detail in the legend to Figure 2 in the paper by Mitchison and Kirschner (45) and thus is not reiterated here.

1. Induce microtubule assembly by incubating isolated centrosomes (at a final concentration of 5×10^6/ml) in PB buffer (80 mM Pipes, pH 6.8, 1 mM EGTA, 1 mM MgCl$_2$, and 1 mM GTP) with purified tubulin (at a final concentration of 1.2 mg/ml) at 37 °C for 10 min.

2. Dilute the mixture by a factor of 25 into PB containing 30 per cent glycerol at 37 °C.

3. Centrifuge the centrosomes with associated microtubule arrays at 28 000 $\times$ g in a swinging-bucket rotor through a cushion of 40 per cent glycerol in PB on to a polylysine-coated circular cover slip placed at the bottom of a modified 15 ml Corex centrifuge tube. The modification required of the

Protocol 7. *Continued*

centrifuge tube is essentially as described by Evans *et al.* (47) Approximately 10^5 centrosomes per cover slip yield an adequate amount for viewing by video DIC microscopy (46).

4. Incubate 12 μl of motor protein in motility buffer (see Section 5.2) with 4 μl of a 20-fold dilution of latex beads in motility buffer containing 10 μM taxol and 5 mM ATP.

5. Place 15 μl of this mixture on a cover slip, surrounded by a circle of vaccum grease with small cover slip fragments embedded in it to act as spacers, and cover this cover slip with the cover slip containing the attached centrosomes from step 3 above. Seal the preparation with valap and assay for movement using video DIC microscopy.

Beads moving along microtubules toward the centrosomal centre correspond to minus-end ('dynein-like') directed movement, while beads moving out to the periphery of the centrosomal microtubule arrays are indicative of kinesin-like movements (plus-end directed). Bidirectional movements of beads on these microtubule arrays are assumed to be due to the presence of at least two different classes of motor proteins (both plus- and minus-end-directed), as initially observed by Vale *et al.* (46).

6.2 Gliding of microtubules containing polarity-revealing probes

For this class of gliding assay, the microtubules of choice are axonemal microtubules isolated from sperm tail flagella. Use of both morphological (48) and kinetic assays (4, 5) has allowed determination that the axonemal microtubules are oriented with their plus ends facing the tip of the flagellum (i.e., distal to the head in sperm or distal to the cell body in the green alga *Chlamydomonas*). *Chlamydomonas* axonemes have the advantage that the axonemes tend to splay apart at their distal tips during the isolation procedures (4, 49); thus the inherent polarity of the *Chlamydomonas* axonemal microtubules is readily apparent. If a source is available, *Chlamydomonas* flagella can be isolated and the endogenous dynein arms removed by salt extraction essentially as described by King *et al.* (50). If *Chlamydomonas* are not available, this assay can also utilize sperm tail flagella which have been incubated with tubulin subunits. Under appropriate conditions the subunits assemble off the plus ends of the axonemal microtubules, providing a visual marker in the form of a 'tuft' of growth, which can be readily discerned with the video DIC microscope. The details of this tuft approach are provided below, adapted essentially from Pryer *et al.* (51), as employed by Paschal and Vallee (17).

Protocol 8. Assay for movement on axonemes

1. Obtain sea urchin axonemes as described in Chapter 7, *Protocol 2*, or *Chlamydomonas* axonemes as described by King *et al.* (50).

2. Extract the axonemes with 0.6 M KCl (50) to remove the dynein arms by thoroughly resuspending a pellet of axonemes in 4–5 ml ice-cold extraction solution (30 mM Tris-HCl, pH 7.5, 5 mM $MgSO_4$, 1 mM DTT, 0.5 mM EDTA, 0.2 mM ATP, and 0.6 M KCl). Incubate the solution of suspended axonemes for 10 min at room temperature. Collect the axonemes by centrifugation at 35 000 × g for 30 min at 4 °C, leaving the extracted dynein in the supernatant.

3. Resuspend the axonemes in gliding buffer (GB: 20 mM Tris HCl, pH 7.6, 50 mM KCl, 5 mM $MgSO_4$, 0.5 mM EDTA, and 2 mM Mg-ATP) to a suitable concentration such that a twofold dilution yields a reasonable number of axonemes on the microscope slide. The trick here is to have a concentration that yields a few axonemes per microscope field so one does not have to search for them, but not so many that the axonemes overlap and get in each other's way.

4. Add 1 part axoneme solution to 1 part purified tubulin dimers at 1 mg/ml tubulin and 1 mM GTP final concentration, and incubate at 37 °C for 5 min. Under these conditions, the tubulin is at or near its critical concentration, and so self-nucleation will not occur at such short incubation times; rather, nucleation will occur preferentially from the plus ends of the axonemal microtubules, producing a short 'tuft' of microtubule growth from one end of the axoneme. If the tubulin concentration is too low, no growth will be obtained over short incubation times. On the other hand, if the tubulin concentration is too high, other problems occur: if the nucleated microtubules are too long, they fold back on the axoneme, making end definition difficult; at high tubulin concentrations the minus end will also begin nucleating microtubules. In theory, this will not be a problem as the plus end will always be longer. However, as both ends elongate, it becomes more difficult to determine which end has the longer growth. In practice, the correct conditions are arrived at empirically; one should aim for short and discernible microtubules growing from one end only.

5. Coat an 18 × 18 mm #0 cover slip with 5 μl of the motor protein solution to be tested, using a humid chamber as described in Section 5.3. Add 10 μl of axoneme/microtubule suspension in GB, mount and seal the preparation with valap, and observe with video DIC microscopy.

The plus ends of the microtubules will be identified by the tuft of tubulin (remember that a shorter, usually more compact, tuft can occur at the minus end as well). If the axoneme glides with the plus end trailing, this indicates plus-end (or kinesin)-driven movement (i.e., the glass slide is 'moving', relatively speaking, toward the plus end of the axoneme). Thus a cytoplasmic, membrane-bound vesicle with kinesin attached would be driven toward the plus end of the microtubule. Conversely, if the axoneme glides with its plus end leading, this indicates minus-end-directed movement.

Readers should be cautioned that this assay has the potential to be misinterpreted, particularly when applied to dynein or dynein-like motor proteins. Possible reasons for this ambiguity have been previously discussed (see refs 52 and 53). Although the reasons for this ambiguous behaviour are not clear, caution is needed in interpreting results obtained with this assay. One way to be relatively sure of the results is to use frayed *Chlamydomonas* axonemes either initially or to confirm results obtained with the tuft assay. Alternatively, two other approaches outlined in the following two sections have been developed to avoid these potential problems.

6.3 Gliding of microtubules with a fluorescent label at the minus end

This assay employs fluorescently labelled tubulin to create short pieces of microtubules that are cross-linked and then used to nucleate the growth of unlabelled subunits, producing polymers with a fluorescent tag at their minus ends. The procedure outlined below is adapted from several reports (54–56).

Protocol 9. Assay for microtubule gliding using minus-end labelled fluorescent microtubules

1. To prepare fluorescent tubulin, react purified tubulin with carboxy-X-rhodamine-N-hydroxysuccinimide ester (X-rh), essentially as described by Gorbsky *et al.* (54).

2. To produce fluorescent microtubule seeds, assemble X-rh labelled tubulin from step 1 at 3 mg/ml in polymerization buffer (PB: 80 mM Pipes, pH 6.8, 1 mM EGTA, and 1 mM $MgCl_2$) containing 30 per cent glycerol, 4 mM $MgCl_2$, and 1 mM GTP for 20 min at 37 °C. Shear the microtubules to uniform length by passing them five times through a blunt end 22 gauge needle.

3. Cross-link the fluorescent microtubule seeds by the addition of 1–2 mM ethylene glycol-bis-succinimidylsuccinate (EGS) in dimethyl sulphoxide (DMSO) and incubation at 37 °C for 15 min.

4. Quench the cross-linking reaction by dilution of the microtubules into 1 ml of PB containing 50 per cent sucrose, 10 mM glutamate, and 0.5 mM 2-mercaptoethanol and incubation at room temperature for 1 h. EGS-

cross-linked microtubules cannot be frozen and thawed, but they are stable for 2–3 months at room temperature (55).

5. Elongate the fluorescent microtubule seeds selectively from their plus ends by incubating them in the presence of NEM-treated tubulin. To do this, incubate 4 mg/ml of purified tubulin with 1 mM N-ethylmaleimide (NEM) for 5 min on ice; inactivate the unreacted NEM with 10 mM DTT for 15 min.

6. React the NEM-treated tubulin at 1.5 mg/ml with the fluorescent microtubule seeds in PB for 30 min at 37 °C. Under these conditions the NEM-tubulin polymerizes preferentially from the plus ends of the fluroescent seeds. After 30 min, dilute 50 μl of this solution into 1 ml of PB containing 20 μM taxol to stabilize the microtubules.

7. Use these microtubules in a gliding assay, as described for the axonemes in the preceding section, with observation by fluorescence and video DIC microscopy.

Because the labelled end of the microtubule composed of fluorescent tubulin subunits has been cross-linked with EGS, one can be reasonably sure the label will remain in place during the course of the experiment. Kinesin will move such an asymmetrically labelled microtubule across the glass surface with its minus (and hence labelled) end leading, while dynein will move the microtubules in the opposite direction with the labelled end trailing. One disadvantage of this assay is the requirement for fluorescence microscopy, and possibly an expensive image-intensified video camera (see Chapter 1), to visualize the fluorescent microtubule ends. Thus, perhaps the ultimate motor assay is described next, in which intact, detergent-extracted sperm with the heads still attached to the tails are used. The presence of the heads not only reveals the polarity of the axonemal microtubules but allows the observation of the gliding movements relatively easily (and inexpensively) with the phase microscope.

6.4 Gliding of demembranated sperm with the head still attached

This straightforward assay is ingenious in its simplicity. Extract the membranes and the dynein from sperm flagellar axonemes, while leaving the heads still attached and intact. Thus, the heads provide the marker for the minus ends of the microtubules. The assay was originally reported by Lye et al. (53) and subsequently used by McDonald et al. (57). The procedure detailed below has been generated by the author from information generously provided by Dr John Lye (Washington University School of Medicine) and by Drs Heather McDonald and Larry Goldstein (Harvard University) from a method provided by Dr Bill Saxton (Indiana University).

Protocol 10. Assay for microtubule gliding using demembranated sperm

1. Collect sea urchin sperm 'dry' by shedding onto a watch glass or other suitable receptacle. The dry sperm can be stored for a few days at 4 °C, but are generally used for this motility assay on the day of collection.

2. Extract the sperm by incubating 1 part dry sperm with 3 parts demembranation/dynein extraction solution, (DDE; 10 mM Tris, pH 8.0, 0.5 M KCl, 2 mM $MgSO_4$, 0.5 mM EDTA, 1 mM DTT plus detergent): the original assay called for 0.04 per cent Triton X-100, which was subsequently increased to 0.1 per cent. This concentration extracts the flagellar membrane well, but can cause some extraction of the membrane of the head as well, thus releasing DNA. John Lye has examined various detergents and has found that 0.1 per cent Brij-58 provides the best extraction of the sperm tails while leaving the sperm heads intact. Mix gently, and incubate at least 20 min at room temperature, mixing occasionally by gentle inversion. Alternatively, one can monitor the process by DIC microscopy to verify extraction of the flagellar membranes. The sperm should stop swimming when the extraction is complete.

3. Dilute the extracted sperm 10-fold with PEM buffer (80 mM Pipes, pH 6.8, 1 mM EGTA, 2 mM $MgCl_2$) containing 1 mM DTT.

4. Adsorb approximately 10 μl of motor protein in buffer containing 2 mM Mg-ATP on to a glass cover slip for 10 min, using a humid chamber as previously described.

5. Add 1 μl diluted, extracted sperm to the cover slip, and allow the sperm to adsorb for another 5–10 min before inverting the cover slip on to a glass slide. Alternatively, a perfusion chamber can be constructed and the extracted and demembranated sperm perfused through the chamber which has been previously coated with the motor protein. This procedure is described briefly in Section 5.3 and in detail in refs 43 and 58.

6. Observe the preparation with video DIC microscopy or more simply with phase microscopy, as the sperm heads and attached axonemes should be readily apparent.

This assay is by far the simplest and most straightforward. No protein purification and assembly steps are required to generate the gliding microtubules; no elaborate procedures are necessary to generate polarity-revealing probes, as the attached heads mark the minus ends of the sperm tail axonemal microtubules. If the sperm glide with their heads trailing, this indicates the presence of a dynein-like, minus-end-directed motor, while if the heads are leading, this indicates a plus-end (or kinesin-like) motor is responsible.

Roger D. Sloboda

7. Concluding remarks

Some general comments about the methods presented in this chapter are perhaps appropriate here.

- The quality of the cover slips used in the assays described here is very important. Different manufacturers apparently clean the glass used in the manufacture of cover slips with various solvents, and presumably residues left over from the cleaning process can inhibit various forms of microtubule-based motility. Thus, in practice a number of laboratories will purchase several boxes of cover slips from different manufacturers, test samples of each in a known motility assay (e.g. kinesin-induced microtubule gliding), and use only those boxes of cover slips that support normal gliding (i.e., at a rate of 0.3–0.5 μm/s, with motility occurring in each of a number of different fields on a given cover slip).

- The gliding assays described here include one step where the presumptive motor protein is adsorbed to the glass cover slip. For a complete description of this process and how it is performed in a gliding assay using axonemal dynein, see the details in Sale and Fox (58). These authors have done a thorough analysis of the procedural variables, and gliding powered by axonemal dynein is therefore very reproducible when using their methods.

- Note that the rates and characteristics of microtubule gliding given here are average values taken from the literature. The reader should be aware, however, that not all preparations produce gliding with the same characteristics, as described by Porter *et al.* (59). For example, these authors have noted that as the active kinesin concentration decreases due to ageing or dilution, gliding becomes progressively more irregular. Eventually, gliding ceases, although microtubule bending and flexing can occur. These authors also note that gliding of axonemes is less consistent than the gliding of single microtubules. This is observed as fewer axonemes gliding per field than in an otherwise identical field composed of single microtubules. Yet Porter *et al.* (59) note that when an axoneme does glide, it moves at a rate comparable to a single microtubule under similar conditions. These characteristics were noted with kinesin-induced gliding, and although the same may or may not be true for dynein-induced gliding, their analysis nevertheless points out the need for trying varying concentrations of motor protein, varying the source of microtubules, etc.

- Finally, the importance of using one of the specific directional gliding assays outlined in Section 6 is best exemplified by the recent results of Walker *et al.* (56) and McDonald *et al.* (57). The N-terminal globular head region of the kinesin heavy chain contains the ATP- and microtubule-binding domains of the intact molecule (60, 61). Sequence analysis of

193

related proteins has led to the discovery of a kinesin-like family of proteins having an average of about 40 per cent identity at the amino acid level to the N-terminal motor domain of kinesin. The *ncd* gene at the *claret* locus in *Drosophila* encodes one of these kinesin-like proteins, but in the case of the *ncd* gene product the motor domain, containing kinesin consensus sequences for ATP and microtubule binding, occurs at the C-terminus of the protein instead of the N-terminus as in normal kinesin (62, 63). When the protein encoded by the *ncd* gene was expressed in bacteria and used in a microtubule gliding assay, it was determined that the protein had minus-end-directed motor activity, similar to dynein, yet the *ncd* gene product motor domain shares 40–45 per cent identity at the amino acid level with the kinesin motor (56, 57). Such a result dramatically emphasizes the need to verify the characteristics of microtubule motor proteins using one of the assays outlined here.

Acknowledgments

The author gratefully acknowledges support from the NSF (DCB-88-20553) and the NIH (GM43982) and thanks Dr John Lye (Washington University School of Medicine) and Drs Heather McDonald and Larry Goldstein (Harvard University) for generously providing details of the sperm gliding procedure outlined in Section 6.4. Thanks are also extended to Drs Mary Lou Guerinot and C. Robertson McClung (Dartmouth College) for their comments and support and to Dr Susan P. Gilbert (Pennsylvania State University) for a critical reading of the manuscript, as trained to do by the Coach.

References

1. Grafstein, B. and Forman, D. S. (1980). *Physiol. Rev.*, **60**, 1167.
2. Schliwa, M. (1984). *Cell Muscle Motil.*, **5**, 1–82.
3. Weisenberg, R. C. (1972). *Science*, **177**, 1104.
4. Binder, L. I., Dentler, W. L., and Rosenbaum, J. L. (1975). *Proc. Natl. Acad. Sci. USA*, **72**, 1122.
5. Bergen, L. G. and Borisy, G. G. (1980). *J. Cell Biol.*, **84**, 141.
6. Kirschner, M. W. (1980). *J. Cell Biol.*, **86**, 330.
7. Mitchison, T. and Kirschner, M. (1984). *Nature*, **312**, 237.
8. Heidemann, S. R. and McIntosh, J. R. (1980). *Nature*, 286, 517.
9. Telzer, B. R. and Haimo, L. T. (1981). *J. Cell Biol.*, **89**, 373.
10. Heidemann, S. R., Landers, J. M., and Hamborg, M. A. (1981). *J. Cell Biol.*, **91**, 661.
11. Euteneuer, U. and McIntosh, J. R. (1981). *J. Cell Biol.*, **89**, 338.
12. Weiss, P. and Hiscoe, H. B. (1948). *J. Exp. Zool.*, **107**, 315.
13. Brady, S. T. and Lasek, R. J. (1982). *Meth. Cell Biol.*, **25**, 365.

Roger D. Sloboda

14. Willard, M., Cowan, W. M., and Vagelos, P. R. (1974). *Proc. Natl. Acad. Sci. USA*, **71**, 2183.
15. Weiss, D. G. (ed.) (1982). *Axoplasmic transport*. Springer-Verlag, New York.
16. Vale, R. D., Reese, T. S., and Sheetz, M. P. (1985). *Cell*, **42**, 39.
17. Paschal, B. M. and Vallee, R. B. (1987). *Nature*, **330**, 181.
18. Nomarski, G. (1955). *J. Phys. Radium*, **16**, 9.
19. Allen, R. D., David, F. B., and Nomarski, G. (1969). *Zeitschr. f. wiss. Mikr. und Mikrotech.*, **69**, 193.
20. Allen, R. D., Allen, N. S., and Travis, J. L. (1981). *Cell Motil.*, **1**, 291.
21. Inoué, S. (1981). *J. Cell Biol.*, **89**, 346.
22. Walter, R. J. and Berns, M. W. (1981). *Proc. Natl. Acad. Sci. USA.*, **78**, 6927.
23. Allen, R. D. and Allen, N. S. (1983). *J. Microscopy*, **129**, 3.
24. Inoué, S. (1986). *Video microscopy*. Plenum Press, New York.
25. Spencer, M. (1982). *Fundamentals of light microscopy*. Cambridge University Press.
26. Gilbert, S. P., Allen, R. D., and Sloboda, R. D. (1985). *Nature*, **315**, 245.
27. Gilbert, S. P. and Sloboda, R. D. (1989). *J. Cell Biol.*, **109**, 2379.
28. Schnapp, B. J. (1986). *Meth.* Enzymol. **134**, 561.
29. Schiff, P. B., Fant, J., and Horwitz, S. B. (1979). *Nature*, **277**, 665.
30. Weingarten, M. D., Lockwood, A. H., Hwo, S.-Y., and Kirschner, M. W. (1975). *Proc. Natl. Acad. Sci. USA*, **72**, 1858.
31. Borisy, G. G., Olmsted, J. B., Macrum, J. M., and Allen, C. (1974). *Fed. Proc.*, **33**, 167.
32. Lasek, R. J. and Brady, S. T. (1985). *Nature*, **316**, 645.
33. Kuznetsov, S. A. and Gelfand, V. I. (1986). *Proc. Natl. Acad. Sci. USA*, **83**, 8530.
34. Bloom, G. S., Wagner, M. C., Pfister, K. K., and Brady, S. T. (1988). *Biochemistry*, **27**, 3409.
35. Hackney, D. D. (1988). *Proc. Natl. Acad. Sci. USA*, **85**, 6314.
36. Paschal, B. M., Shpetner, H. S., and Vallee, R. B. (1987). *J. Cell Biol.*, **105**, 1273.
37. Porter, M. E., Scholey, J. M., Stemple, D. K., Vigers, G. P. A., Vale, R. D., Sheetz, M. P., and McIntosh, J. R. (1987). *J. Biol. Chem.*, **262**, 2794.
38. Summers, K. E. and Gibbons, I. R. (1971). *Proc. Natl. Acad. Sci. USA*, **68**, 3092.
39. McGee-Russell, S. M. and Allen, R. D. (1971). *Adv. Cell Mol. Biol.*, **1**, 153.
40. Sheetz, M. P. and Spudich, J. A. (1983). *Nature*, **303**, 31.
41. Vale, R. D., Schnapp, B. J., Reese, T. S., and Sheetz, M. P. (1985). *Cell*, **40**, 559.
42. Allen, R. D., Weiss, D. G., Hayden, J. H., Brown, D. T., Fujiwake, H., and Simpson, M. (1985). *J. Cell Biol.*, **100**, 1736.
43. Vale, R. D. and Toyoshima, Y. Y. (1988). *Cell*, **52**, 459.
44. McIntosh, J. R. and Porter, M. E. (1989). *J. Biol. Chem.*, **264**, 6001.
45. Mitchison, T. and Kirschner, M. (1984). *Nature*, **312**, 232.
46. Vale, R. D., Schnapp, B. J., Mitchison, T., Steuer, E., Reese, T. S., and Sheetz, M. P. (1985). *Cell*, **43**, 623.
47. Evans, L., Mitchison, T., and Kirshner, M. (1985). *J. Cell Biol.*, **100**, 1185.
48. Haimo, L. T., Telzer, B. R., and Rosenbaum, J. L. (1979). *Proc. Natl. Acad. Sci. USA*, **76**, 5759.

49. Allen, C. and Borisy, G. G. (1974). *J. Mol. Biol.*, **90**, 381.
50. King, S. M., Otter, T., and Witman, G. B. (1986). *Meth. Enzymol.*, **134**, 291.
51. Pryer, N. K., Wadsworth, P. W., and Salmon, E. D. (1986). *Cell Motil. Cytoskeleton*, **6**, 537.
52. Lye, R. J., Porter, M. E., Scholey, J. M., and McIntosh, J. R. (1987). *Cell*, **51**, 309.
53. Lye, R. J., Pfarr, C. M., and Porter, M. E. (1989). In *Cell movement*, Volume 2: *Kinesin, dynein, and microtubule dynamics*, ed. F. D. Warner and J. R. McIntosh, pp. 141–154. Alan R. Liss, New York.
54. Gorbsky, G. G., Sammak, P. J., and Borisy, G. G. (1988). *J. Cell Biol.*, **106**, 185.
55. Koshland, D. E., Mitchison, T. J., and Kirschner, M. W. (1988). *Nature*, **331**, 499.
56. Walker, R. A., Salmon, E. D., and Endow, S. A. (1990). *Nature*, **347**, 780.
57. McDonald, H. B., Stewart, R. J., and Goldstein, L. S. B. (1990). *Cell*, **63**, 1159.
58. Sale, W. S. and Fox, L. A. (1988). *J. Cell Biol.*, **107**, 1793.
59. Porter, M. E., Grissom, P. M., Scholey, J. M., Salmon, E. D., and McIntosh, J. R. (1988). *J. Biol. Chem.*, **263**, 6759.
60. Yang, J. T., Laymon, R. A., and Goldstein, L. S. B. (1989). *Cell*, **56**, 879.
61. Yang, J. T., Saxton, W. M., Stewart, R. J., Raff, E. C., and Goldstein, L. S. B. (1990). *Science*, **249**, 42.
62. Endow, S. A., Henikoff, S., and Soler-Niedziela (1990). *Nature*, **345**, 81.
63. McDonald, H. B. and Goldstein, L. S. B. (1990). *Cell*, **61**, 991.

9

Methods for studying the cytoskeleton in yeast

F. SOLOMON, L. CONNELL, D. KIRKPATRICK, V. PRAITIS,
and B. WEINSTEIN

1. Introduction

Yeast offers obvious advantages for studying basic processes in cell biology and physiology at the molecular level. Because of the excellent genetics available in yeast, combined with powerful molecular techniques, several issues are being successfully attacked. These include metabolism, transcription, mobile genetic elements, protein metabolism and processing, and even aspects of differentiation. The information obtained from these studies, as well as the sorts of approaches used, are applicable to higher eukaryotes, attesting to the power of yeast as a model system.

It is perhaps less obvious that yeast should become such a useful system for analysing cytoskeletal structure and function, since it does not display some of the more dramatic of those functions, such as walking across substrata, or extending axons. However, yeast cells do contain the cytoskeletal structures, microtubules and microfilaments, required for cell motility in all eukaryotes, and the genes encoding the major components of those structures are essential for cell viability. Mutations in these elements affect fundamental processes such as cell division, chromosome segregation, and expression of cellular polarity. Therefore, working out the genetics of the cytoskeleton in yeast is both feasible and attractive. The genetic analyses are supported by cytological and biochemical methods developed in recent years. The rationales and results of these approaches have been reviewed extensively (1–4). Several groups now use yeast as their major tool for analysing the cytoskeleton. Other groups are drawn to yeast in search of potentially interesting homologies to proteins in higher eukaryotes, or more precise *in vivo* assays for a particular gene product. This chapter is directed toward the latter group. We describe here the basic methodologies which are needed routinely in studying the yeast cytoskeleton. These include not only experimental strategies such as identification of interacting genes, but also solutions to more general problems, such as protecting proteins against yeast proteases. Several are procedures which we have been taught by the many helpful laboratories which form the yeast cell biology community; they are reported here with our modifications. Throughout, we have supplied references where more detailed treatments of these topics can be obtained.

2. Genetic analysis of mutants

2.1 Transforming yeast strains

Yeast cells are readily transformed with DNA. The transformants can integrate the foreign DNA into their chromosomes, usually by site-specific integration events, or they can carry the DNA on extrachromosomal plasmids appropriately constructed to replicate and even to segregate. *Protocols 1* and *2* below describe procedures for transformation and comment on routine maintenance and characterization of transformants. All media are prepared according to Sherman *et al.* (5).

2.1.1 Transformation protocols

Two popular protocols for transforming yeast strains with DNA are described in detail below. Different protocols work better for different yeast strains, even within the same laboratory. The DNA can come from *E. coli* 'mini' preparations, columns, or caesium chloride gradients.

Protocol 1. Calcium transformation (modified from ref. 6)

Reagents

- TE-Buffer: 10 mM Tris, 1 mM EDTA, pH 7.8–8.0 (adjust with concentrated HCl); autoclave.
- CATE: 10 mM $CaCl_2$ in TE; autoclave.
- PEG4000: 4.0 g polyethylene glycol 4000, 7.0 ml TE; filter sterilize.
- SOS: 1 M sorbitol, 10 mM $CaCl_2$, 33 per cent liquid YPD (5); autoclave.

Procedure

1. Seed an overnight culture in selection medium.
2. Inoculate 100 ml of YPD in a 500 ml flask, using the overnight culture. Place at 30 °C with shaking.
3. Harvest cells when $A_{600} = 0.2$ by centrifuging for 5 min at 2500 × g. All centrifugations are for 5 min at 2500 × g unless otherwise indicated.
4. Wash once with 10 ml TE buffer. Centrifuge.
5. Resuspend in 5 ml CATE buffer. Transfer to a 250 ml sterile Erlenmeyer flask. Gently agitate at 30 °C for 30 min. Centrifuge.
6. Resuspend cells in 0.5 ml TE. Divide into 0.1 ml aliquots in small test-tubes.

7. Add 1–5 µg DNA (caesium chloride-purified) to each tube. Incubate at room temperature for 10 min.

8. Heat-shock in 42 °C water bath for 5 min.

9. Allow tubes to sit at room temperature for 10 min.

10. Add 1 ml 40 per cent PEG4000 to each tube.

11. Allow tubes to sit at room temperature for 45 min. Centrifuge.

12. Resuspend in 0.5 ml SOS.

13. Spread-plate 0.1–0.25 ml of SOS resuspension on selection plates.

14. Place plates at 30 °C for 2–3 days.

Protocol 2. Lithium transformation (modified from ref. 7)

Reagents

- 10× TE-Buffer: 0.1M Tris-HCl, 0.01M EDTA, adjust pH to 7.5 with HCl, filter-sterilize.

- 10× LiAc: 1 M LiAc (Sigma L-6883); filter-sterilize.

- 50 per cent PEG3350: (w/v) (Sigma P-3640); filter-sterilize.

- Carrier DNA: Calf thymus DNA (Sigma) suspended in distilled water to 3 mg/ml and violently sonicated (with a probe sonicator) until it is no longer viscous. Fragments of a few thousand base pairs are optimum. Extract with phenol–chloroform–isoamyl alcohol (24:24:1) till clean, precipitate with ethanol and resuspend to about 3 mg/ml.

Procedure

1. Seed an overnight culture.

2. Inoculate 50 ml growth media with approximately 10 ml from the overnight culture.

3. When the density reaches 2×10^7 cells/ml, gently pellet the cells at 1000 rev/min (IEC, small rotor) for 5–10 min.

4. Resuspend in 2.5 ml 1 × TE/0.1 M LiAc; pellet. Resuspend in 0.5 ml 1× TE/0.1 M LiAc; incubate 1 h at 30 °C.

5. Prepare sterile Eppendorf tubes containing plasmid DNA (0.5–1 µg) of integrating fragment works best) and 45 µg of carrier DNA. At the end of the 1 h incubation, gently mix the cells to suspend uniformly and add 100 µl cells to each tube, pipetting up and down a few of times to mix. Incubate 0.5 h at 30 °C.

6. Add 0.7 ml 40 per cent PEG/0.1 M LiAc in TE to each tube. Vortex briefly and incubate 1 h at 30 °C. The suspension may appear flocculent. Pre-warm a bath to 42 °C to have available in 1 h.

Protocol 2. *Continued*

7. Incubate 5 min at 42 °C. Pellet the cells 2 sec in a microfuge (12 000 × *g*). Wash twice in 1× TE (resuspend by pipetting up and down). Resuspend the pellets in 500 µl TE and spread 250 µl on each of two plates.

● To maximize transformation efficiency when doing several transformations at once, wash the cells in SD or SCD medium minus the selective marker, instead of in TE.

2.1.2 Some remarks on handling yeast transformants

In general, it is necessary to analyse several independent transformants. Typically, since transformation itself is mutagenic, there are some abnormal diploid transformants which fail to sporulate or do not give 2:2 segregation of markers, or, in the case of haploids, show incorrect integration or abnormal segregation, as assesed by Southern blots or other criteria.

There are various schools of thought regarding how extensively to grow stocks before and during characterization. Classical genetic technique calls for streaking to obtain strains derived from single cells. In situations where complications such as reversion, gene conversion, or appearance of unlinked suppressors may occur, it may be advisable to patch directly from the transformation plate, freeze from the patch, and characterize the strains with the minimum of cell divisions needed to generate enough cells. A middle course involves streaking isolates from the transformation plate, then picking a single average-sized colony from that streak to grow cells for freezing. From frozen stock, streak on to a plate, and pool half a dozen or so average-size colonies when seeding cultures for experiments. Alternatively, one can seed liquid cultures directly from the frozen stock. Ideally, frozen stocks should give colonies of uniform size when streaked from the freezer, but even some presumed wild-type strains can give a lot of variability. When characterizing transformants, freeze candidates as soon as possible and then later discard ones that behave abnormally (fail to sporulate, etc.). If one waits until after the sporulation and dissection to freeze, concerns arise about changes incurred while the cells were sitting on plates. Maintaining plates in the cold is not advisable, since cold is reported to activate transposable elements. Similarly, cells sitting on plates for a long time at room temperature can change. Strains can be carried by patching on plates, repatching at 4–5 day intervals, and then streaking from each patch, with no problems. However, one should be alert for changes in the uniform size and normal appearance of the colonies.

2.1.3 Growth and viability assays (8, 9)

Diploid *Saccharomyces cerevisiae* can be induced to undergo meiosis and produce spores when starved for nitrogen on a non-fermentable carbon source such as acetate. The four products of meiosis, the haploid spores, are contained within the mother cell, or ascus. This ascus can be removed by digestion with enzymes such as zymolyase or glusulase, and the four spores removed by micro-manipulation. Typically, each spore is then positioned at a defined position on solid media and allowed to grow. The four colonies of haploid cells that result from a single ascus, and therefore are the meiotic products of a single cell, are collectively called a tetrad. When tetrads are plated on solid media lacking various amino acids, any cells auxotrophic for that amino acid will fail to grow. For example, a diploid cell mutant in one copy of the URA3 gene, encoding orotidine-5-phosphate decarboxylase, will give rise to four colonies after sporulation, two of which will fail to grow on media lacking uracil. If two loci were mutant in the original diploid, the frequency of recombination between the two genes can be determined, allowing the mutations to be mapped relative to one another. Strains exist which have multiple auxotrophies on various chromosomes, allowing new mutations and genes to be quickly mapped. The gene to be mapped is disrupted *in vitro* through insertion of a marker gene. This construct is then integrated into the chromosome by homologous recombination in a diploid strain mutant in both copies of the marker gene. Then, after sporulation the marker gene is examined for linkage to genes which have already been mapped. Tetrad analysis can also be used to determine if a mutation, originally isolated in a diploid, is an essential gene. If the gene that is mutated is essential, only two spores will grow after sporulation, given that the mutation produces a non-functional protein.

2.2 Generation of mutants

2.2.1 Selection for drug resistance: benomyl

The drug benomyl and its derivatives are known to interact with tubulin *in vitro*, and to depolymerize microtubules *in vivo* (5,10–13). The drug itself is lethal at relatively low doses. Consequently, mutations in genes encoding microtubule components, or affecting microtubule functions, have been identified by screening for resistance or hypersensitivity to benomyl. In addition, benomyl sensitivity assays are used as phenotypic screens for mutations in microtubule proteins. See *Protocol 3*.

Protocol 3. Selection of mutants with benomyl

Reagents

● Benomyl: 10 mg/ml stock in dimethyl sulphoxide; store at 4 °C.

Procedures

1. Make media and maintain at 60 °C in water bath.

2. Add benomyl to appropriate concentrations in 200 μl aliquots, stirring vigorously after each addition.

Protocol 3. *Continued*

3. For increased sensitivity, apply a range of cell concentrations to each benomyl plate. 50 μg/ml of benomyl is an effective concentration for wild-type cells at room temperature, but lower concentrations are required for cells grown at lower temperatures.

2.2.2 Generation of mutants by *in vitro* mutagenesis

In the absence of a genetic or drug screen for a desired phenotype, chemical or radiation-induced mutants can be generated and screened individually for cytoskeletal defects. This is, however, a tedious and time-consuming task. If a cytoskeletal gene is in hand, mutations can be generated easily *in vitro* using methods such as misincorporation mutagenesis or hydroxylamine mutagenesis. The changes can then be introduced into the genome through transformation by use of an integrating plasmid or via the plasmid shuffle method (14–17).

2.2.3 Integration by site-specific recombination

Linear DNA fragments recombine with homologous sites in the chromosome when transformed into *Saccharomyces cerevisiae*. This process allows the introduction into the chromosome of mutations constructed *in vitro*. The integration events can be directed such that the mutant protein is produced utilizing the 5′ non-coding region of the chromosomal version of the gene. To accomplish this, the desired mutation is introduced into the coding region of a plasmid-borne gene of interest *in vitro* and confirmed by sequencing. Then, the 5′ non-coding region and the sequence corresponding to the first few amino terminal amino acids are removed from the gene *in vitro*. Finally, the plasmid is introduced by transformation into yeast, selecting for the auxotrophic marker on the plasmid. The plasmid itself should be incapable of self-replication in yeast, so that only transformation events that result in integration are detected. Thus, any transformants bearing the auxotrophic marker should have integrated the plasmid into the genome. Alternatively, a linear fragment, containing the mutated gene of interest and the selectable marker, can be transformed into the genome. Free linear ends of DNA fragments are very recombinogenic in yeast. Linearizing near the 5′ end of the gene of interest will efficiently direct integration to that region. The effect of either of these transformations is to replace the chromosomal copy of the gene with the introduced copy. Thus, it is possible to assess the *in vivo* consequences of *in vitro* constructed mutations in yeast, even when they are recessive (e.g., ref. 18).

2.2.4 Introduction of mutants on extrachromosomal DNA (plasmid shuffle)

The 'plasmid shuffle' technique (*Protocol 4*) can be used for replacing the chromosomal wild-type copy of a gene with a mutant copy on a plasmid

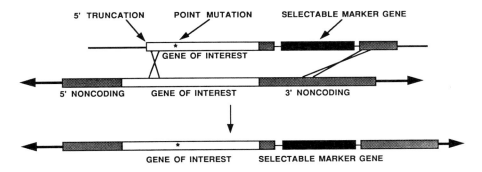

Figure 1. Integration of mutants via site-specific recombination. A truncated version of the gene of interest containing a point mutation (indicated by *) and a selectable marker in the 3' non-coding region of the gene is allowed to recombine with the chromosomal copy of the gene of interest. The resulting chromosomal recombinant possesses both the point mutation and the selectable marker.

(*Figure 2*). The mutant can be either a single mutation generated by site-directed mutagenesis or a library of mutants in the gene of interest generated by chemical methods. This procedure is designed for use with an essential gene that is not sensitive to moderate overexpression of the gene product (17, 19).

Protocol 4. Plasmid shuffle technique

1. Place the wild-type copy of the gene of interest on a plasmid containing the yeast URA3 gene as a marker.

2. Transform a strain which is Ura⁻ and contains at least two other auxotrophies, such as His⁻ and Leu⁻, with the above plasmid, selecting for Ura⁺ colonies.

3. Transform the Ura⁺ strain with an integrating plasmid or a linear piece of DNA containing a disrupted or disrupted and deleted copy of your gene, selecting for the marker, such as His⁺, associated with the disruption. There are three possible His⁺ transformants, depending upon the site of integration of the disruption:

   • Integration at the chromosomal gene of interest. This is the desired result, with the URA3 plasmid stably maintained even under non-selective conditions.

   • Integration into the URA3 plasmid itself at the gene of interest. With an integration into the plasmid, neither the Ura⁺ marker nor the His⁺ marker will be maintained under non-selective conditions.

Protocol 4. *Continued*

- Integration at the site of the chromosomal HIS gene. In this case, the URA3 plasmid will not be maintained under non-selective conditions, but the His$^+$ marker will be maintained.

4. Generate another plasmid with a third marker (Leu$^+$) and containing a mutagenized copy of your gene of interest.

5. Transform and select for Leu$^+$, without selecting for Ura$^+$.

6. Replica-plate colonies to plates containing 5-fluoro-orotic acid (5-FOA) (0.5 mg/ml for 37 °C plates, 1.0 mg/ml for 14 °C or 26 °C plates) and to plates without 5-FOA. Colonies that have a functional URA3 gene will not be able to grow on 5-FOA plates, so this procedure selects for cells which have lost the URA3 plasmid.

7. Take colonies from 5-FOA plates and check for growth on 5-FOA plates under non-permissive conditions such as 14 °C or 37 °C. Plasmids with possible conditional alleles of the gene should be isolated and taken through this procedure again to verify the ability of the plasmid to confer a conditional phenotype.

2.3 Suppression analysis

An important goal in analysis of cytoskeletal structures is the identification of proteins which interact *in vivo* to build structures or to regulate their assembly or function. The genetic means for identifying such gene products is suppression analysis. Beginning with the original conditional mutation as background, one can design screens for alterations in other genes which produce a wild-type phenotype. The approaches described below, second site reversion, suppression by overexpression, and non-complementation, are in fact quite complicated, and typically require sophisticated genetic treatments which are beyond the scope of this book. However, it is possible to state the principle behind each of these paradigms, and to describe in some detail how such a suppressor screen can be designed.

2.3.1 Suppression by identification of second site revertants

Jarvik and Botstein (20) demonstrated in the phage P22 that temperature-sensitive mutations can give rise to extragenic revertants which can confer their own conditional phenotype when removed from the original mutant backgound. The extragenic revertants obtained in this manner often identify elements that interact with the product of the gene marked by the original mutation. These temperature-sensitive mutations can themselves be reverted, with the aim of isolating further conditional revertants and extending the chain of interactions.

   Second-site reversion studies have been conducted successfully in yeast

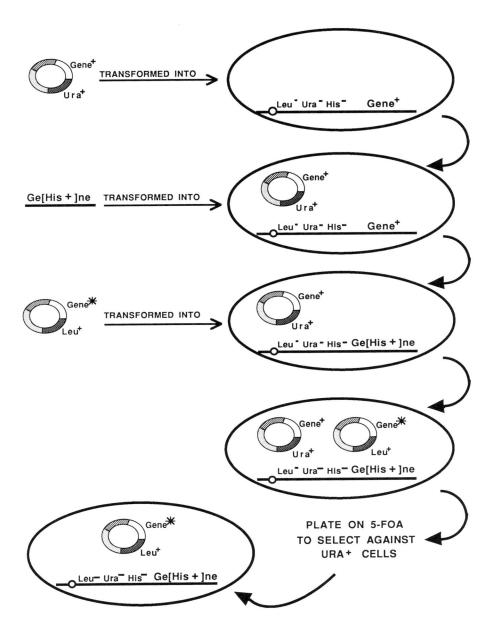

Figure 2. Introduction of mutants on extrachromosomal elements. The wild-type gene of interest is placed on a plasmid containing the URA3 gene as a selectable marker, and this plasmid is transformed into yeast. The chromosomal copy of the gene of interest is then disrupted by integration. A second plasmid, bearing a mutated copy of the gene of interest and another, non-URA3, selectable marker, is then transformed into the strain. Finally, the resulting transformants are plated on 5-fluoro-orotic acid (5-FOA) plates, with selection against cells containing a URA3 gene.

205

(e.g. ref. 21). The general procedure is basically the same as that developed for P22. A colony from a conditionally mutant diploid strain is resuspended and plated at various densities. These plates are incubated at the restrictive temperature. Colonies that form, having presumably acquired a suppressing mutation, are picked, one from each plate to ensure independent events, and patched to a new plate. These colonies can be checked immediately for the gain of a new dominant conditional phenotype by assaying growth under a range of conditions. Alternatively, the strains can be sporulated and the resulting haploids checked for a new, recessive, conditional phenotype. The new mutations can be either intragenic or extragenic, but these possibilities can be distinguished by genetically mapping the site of the mutations in question.

2.3.2 Suppression by overexpression of wild-type genes

A method for identifying genes that produce proteins which interact with your gene of interest is to suppress a mutation in your gene by overexpressing another gene product (*Figure 3* and *Protocol 5*). In yeast this means transforming the mutant strain with a plasmid library of genomic DNA, much as one would do when attempting to clone the gene itself. However, in this case, one looks for strains that suppress the mutant phenotype and yet are not a clone of the gene itself. The level of gene overexpression is dependent on the plasmid vector used for the library. For example, a plasmid containing a centromere will have a copy number of 1 or 2, while a 2-μm plasmid will have a much higher copy number, and therefore a concomitantly higher level of expression of any genes located on the random genomic DNA fragment. For examples of high copy overexpression suppression, see refs. 25 and 26.

Protocol 5. Identifying genes by suppression analysis: overexpression of wild-type genes

1. Identify an allele of the gene concerned with a conditional phenotype, such as a temperature sensitivity, or a scorable phenotype, such as an altered cytoskeletal morphology.

2. Obtain a library, making sure that the mutant strain is auxotrophic for the marker gene on the library plasmid. The library can be either high copy (22, 23) or low copy (24): plasmids containing a yeast centromere (CEN) will be low copy, whereas plasmids containing 2-μm sequences will have a higher copy number.

3. Transform the mutant strain with the library.

4. Either select the transformants for the auxotrophy marker and screen for the mutation, or select for both the auxotrophy marker and the mutation. Selecting for both the auxotrophy marker and the mutation allows the isolation of plasmids containing genes which suppress under restrictive

conditions but which are not tolerated in excess copy under non-restrictive conditions.

5. Identify suppressed strains by the loss of the mutant phenotype.
6. Verify that the suppression co-segregates with the plasmid by allowing the strains to lose the plasmid, as assayed by loss of the marker gene, and showing that the mutant phenotype is restored upon loss of the plasmid.
7. Isolate the plasmid, verifying that it is not just a clone of the mutant gene.
8. Retransform the suppressing plasmid into the mutant strain to verify the suppression of the mutant phenotype.

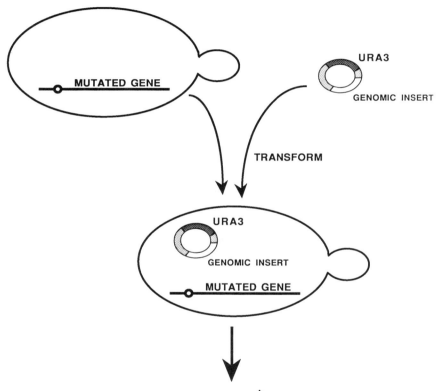

Figure 3. Suppression by overexpression of wild-type genes. A strain containing a conditional mutation in the gene of interest is transformed with a plasmid bearing a selectable marker and a random fragment of genomic DNA. The transformants are then checked for alterations in the conditional phenotype of the mutant.

2.3.3 Suppression by non-complementation

An approach to identifying cytoskeletal proteins involves isolating recessive mutations that fail to complement a second recessive mutation, yet do not occur within that gene (27). Genes encoding proteins that interact closely, as many cytoskeletal elements do, may not be able to complement each other when both are mutant, and such non-complementation should be allele specific. See *Protocol 6*.

Protocol 6. Identifying genes by suppression analysis: non-complementation

1. Obtain a haploid strain containing a conditional lethal mutation in the gene of interest.

2. Mutagenize another haploid strain of opposite mating type and possessing sufficient auxotrophic markers such that when mated to the conditional mutant strain, the resulting diploid can be directly selected.

3. Mate the mutagenized strains to the conditional mutant strain and, in parallel, to a strain isogenic to the mutant but wild-type at that locus.

4. Check the conditional phenotype of the diploids; unlinked non-comple-menters should be able to grow under the restrictive condition when mated with the wild-type strain, but not when mated to the conditional mutant strain.

5. Verify that non-complementers are not linked to the original mutation by tetrad analysis.

---

3. Phenotypic analysis

Analysis of cytoskeletal mutants typically requires both standard yeast methodologies as well as methods for analysing proteins and the structures in which they are found. In this section we describe the techniques for assaying DNA structure and message levels and sizes used commonly and successfully by several groups. We also describe techniques analogous to those used commonly on higher eukaryotes for studying cytoskeletal proteins. In particular, methods for preparing protein samples from yeast are included, with special attention paid to the problem of proteolytic digestion, a serious problem in these cells. A method for fractionating yeast cells to separate 'soluble' and 'cytoskeletal' fractions is included. This particular fractionation preserves both microtubule and F-actin structures. Finally, we describe methods for preparing peptide antibodies and for using antibodies to visualize cytoskeletal structures by immunofluorescence. We have had to omit more specialized assays which may be relevant to some investigations, such as measurements of secretion or rate of chromosome loss. These assays are exhaustively described in current publications, and, in fact, are evolving constantly.

3.1 Protein

Total protein extracts may be prepared from yeast cells in a number of ways. Two of the more convenient methods are glass bead smashing (*Protocol 8*) and French pressing (*Protocol 9*). The first of these methods does not require special equipment and can be used to process a relatively large number of samples in parallel. The second method requires a relatively expensive French Pressure apparatus, and samples must be prepared sequentially. However, yeast cells are more completely disrupted by this procedure and protein denaturation is minimized. General procedures for protein preparation are given in *Protocol 7*.

Protocol 7. Protein preparations: general procedures

Reagents

- GSD (12 ml): 0.62 g dithiothreitol, double-distilled water to 4 ml, 4 ml glycerol, 4 ml 20 per cent SDS. Add a little Bromophenol Blue, using the tip of a metal spatula. Mix well, then let insoluble material settle out. Aliquot convenient amounts and store at −20 °C.

- Protease inhibitors:
  - aprotinin: Comes in solution. Store at 4 °C.
  - PMSF: Make 0.2 M stock in DMSO (49.6 mg in 1.5 ml).
  - TPCK: Make 1 mg/ml stock in DMSO.
  - Leupeptin: Make 1 mg/ml stock in DMSO.
  - Pepstatin: Make 1 mg/ml stock in DMSO.

All these inhibitors are available from Sigma Chemical Company. For all except aprotinin, aliquot convenient volumes (20 μl) and store at −20 °C; thaw and dilute just before use.

Procedure

1. Prepare in advance labelled tubes for smashing cells (if using the glass bead smash, *Protocol 8*) and for aliquoting the final product (both protocols). Immediately before starting, take some GSD out of the freezer to thaw and begin a boiling water bath. If using the French Press method (*Protocol 9*), put the pressure cell on ice to chill it. Note: It is important to minimize delays between harvesting the cells and freezing the final product, as protein degradation can occur even in the presence of the protease inhibitors.

Protocol 7. *Continued*

2. Harvest cells in mid-log phase ($1-2 \times 10^7$ cells/ml culture). Convenient amounts are approximately 2×10^8 cells for glass bead smashing and 2×10^9 cells for French Pressing. Centrifuge at $2500 \times g$ at room temperature (or culture temperature) for 6–8 min.

3. While cells are centrifuging, add 10 µl of each protease inhibitor to 10 ml PBS while vortexing. All of the inhibitors, PMSF in particular, tend to come out of solution unless the PBS is vortexed throughout the addition process.

4. Use either the glass bead smash (*Protocol 8*) or the French Press method (*Protocol 9*) to lyse the cells.

Protocol 8. Lysis by glass bead smash

Materials

● Glass beads: Approximately 0.4 mm diameter beads. Wash for 1 h in concentrated nitric acid and then rinse thoroughly in double-distilled water, dry at 37 °C overnight, then bake in a glassware oven.

Procedures

1. Resuspend cells from *Protocol 7* in 1 ml PBS (without protease inhibitors), transfer to Eppendorf tubes and centrifuge for 4 s. Remove the supernatant. Add glass beads to approximately the same volume as the cell pellet. Add 150 µl PBS with protease inhibitors. Use pipette tip to mix the beads into the cell pellet. Put on ice.

2. Vortex at top speed for 30 s, then put in ice for 30 s. Repeat these stages four more times.

3. Add more PBS with protease inhibitors (optional—see note below). Add ½ volume GSD and mix it in briefly.

4. Place in a boiling water bath for 5 min.

5. Centrifuge for 20 s.

6. Aliquot convenient volumes and freeze immediately on dry ice. Store at −80 °C.

Notes

● You may find it convenient to dilute the extract in stage 3. For samples to be run on standard-size gels, add another 450 µl PBS with protease inhibitors, making a total of 600 µl, and then 300 µl GSD. Approximately 20 µl of this is a moderate load for tubulin analysis.

- If native protein is required, the cleared extract can be used immediately before adding the GSD.
- Scaling up for larger harvests: For 1×10^9 cells, smash in 300 µl PBS. Yields may decrease as pellet size increases.

Protocol 9. Lysis by French press

1. Resuspend cells from *Protocol 7* in 10 ml PBS to wash the cells, and centrifuge again.
2. Pour off the supernatant and resuspend the pellet in approximately 2 ml of PBSA with protease inhibitors. Load the resuspended cell sample into the pre-chilled French pressure cell.
3. Press the sample through the cell at a flow rate of approximately 1 drop/s into a tube on ice. Use a cell pressure of 138 MPa (20 000 p.s.i.) (be sure that the cell you are using is rated to at least this pressure. With an SLM Instruments/Aminco French Press and a 1 cm (⅜ inch) piston diameter pressure cell, the gauge pressure should read 1000 on the *medium* setting). After the sample has been pressed completely through, reload it into the cell and repeat the pressing. More than 95 per cent of the cells are lysed after two passages through the press.
4. Clarify the sample by centrifugation if desired. Add 1–2 volume of GSD, boil an aliquot, as above.

3.1.2 Immunoblots

Several protocols for Western blots of proteins are available, and work well in our hands (28–32). In some circumstances, blots of yeast proteins give lower backgrounds using the TNT method described in *Protocol 10*.

Protocol 10. Immunoblots

Reagents

- 10 × TNT: 0.5 per cent Tween-20; 1.7 M NaCl; 250 mM Tris, pH 7.5.
- PBS: 60.0 g NaCl; 1.50 g KCl; 1.50 g KH_2PO_4; 8.62 g Na_2HPO_4; bring to a total of 6 l with distilled water.

Procedure

1. Dilute antisera in an appropriate volume of TNT. Add to filter, and incubate at room temperature on a rocker platform from 2 h to overnight.
2. Remove diluted antisera and save (they can generally be re-used several times).

Protocol 10. *Continued*

3. Wash filter five times for 15 min each in TNT at room temperature. The volumes of these and subsequent washes are equal to or greater than the volume of the diluted antisera.

4. Dilute the second antibody to the appropriate dilution recommended by the manufacturer. Add diluted antibody to the filter; incubate for 1 h at room temperature on the rocker.

5. Wash five times for 15 min (or longer) each with TNT.

6. Wash twice briefly ($\leq$ 5 min) with PBS.

7. Let filters air-dry on 3MM paper, then mount on cardboard and expose to autoradiographic film, if using radioactive second antibody, or develop colour, if using colour-reactive second antibody (see Chapter 6).

3.2 Nucleic acids

3.2.1 Genomic DNA

The method shown in *Protocol 11* (from ref. 33) can be used for DNA for Southern blots and for genomic libraries. It can also be used to recover plasmids from yeast by subsequent transformation into *E. coli* and re-isolation of the plasmid from the transformed *E. coli*.

Protocol 11. DNA isolation

Reagents

- SCE: 1 M sorbitol, 0.1 M sodium citrate or phosphate (pH 7), 0.06 M EDTA, adjust pH to 7.0.

- GES: 4.5 M guanidine hydrochloride (99 per cent grade), 0.15 M sodium chloride, 0.1 M EDTA, pH 8.0.

Procedure

1. Pellet $10^8–10^9$ cells in the logarithmic growth phase.

2. Resuspend the pellet in 0.3 ml SCE; transfer to a microfuge tube. Add 2 µl β-mercaptoethanol and 20 µl 3 mg/ml Zymolyase 60 000 or 100 000 in SCE. Incubate 20–30 min at 37 °C. Check spheroplasting (loss of cell wall) under the microscope.

3. Pellet in microcentrifuge 2–4 s; drain the pellet. The pellet can be stored frozen at −70 °C.

4. Resuspend in 0.3 ml GES. Add 15 µl of 1 per cent Sarkosyl; incubate at 65 °C for 10 min, mixing occasionally. Cool on ice.

5. Add 0.3 ml cold 95 per cent ethanol and mix well. Pellet 3–5 min in a microcentrifuge at 4 °C; drain pellet well.

6. Using a plastic Eppendorf tip, stir the pellet, then resuspend in 0.6 ml of 10× TE. Add 10 µg RNase and incubate at 65 °C for 15 min. Vortex and incubate an additional 15 min at 65 °C.

7. Add 50 µg proteinase K and incubate for 1 h at 65 °C.

8. Extract with 0.3 ml buffer-saturated phenol and 0.3 ml chloroform/ isoamyl alcohol (24:1 v/v). Centrifuge in microfuge.

9. Remove the aqueous phase to another microcentrifuge tube; repeat the extraction and centrifuge.

10. Remove the aqueous phase to a new tube. Add 0.1 volume 3 M NaOAc and 1 ml 95 per cent ethanol. Precipitate DNA at −20 °C for at least 1 h.

11. Pellet at 4 °C. Drain the pellet and dry. Resuspend in 50 µl 1× TE.

3.2.2 Southern blots

There are several methods for Southern blotting (34, 35). We have found that the sodium hydroxide transfer method (36, 37) works well on the smaller yeast genome, and requires less time and reagents. See *Protocol 12*.

Protocol 12. NaOH transfer method for Southern blotting

1. For fragments greater than 4 kb, depurinate the DNA by soaking in 0.25 M HCl for 10–15 min.

2. Cut four sheets of Whatman 3MM paper so that they overhang the bottom of the gel tray by 5 cm on each end. Pre-wet in distilled water.

3. Place a sponge into a deep dish containing 0.4 M NaOH. Soak the 3MM paper in 0.4 M NaOH and place on top of the sponge. Remove bubbles by rolling a test tube over the blotting paper.

4. Add more 0.4 M NaOH to the paper, and carefully place the gel on it. Remove any trapped air bubbles and cover the gel with a small amount of NaOH.

5. Place plastic wrap over the entire apparatus. Cut a window with a razor blade such that only the gel is exposed.

6. Pre-wet a nylon membrane and lower onto the gel surface. Flood the surface with NaOH and remove air bubbles. Place two more sheets of pre-wetted 3MM blotting paper, cut to the gel size, on top of the nylon. Be careful to remove bubbles.

7. Stack paper towels on top of the 3MM paper and cover with a glass plate.

8. Transfer for 2–24 h, depending on the gel concentration and fragment size.

Protocol 12. *Continued*

9. After transfer, carefully remove the nylon membrane and rinse in the pre-hybridization buffer of choice. Let the membrane air dry. This blot can be stored at 4 °C, wrapped tightly, for several months.

3.2.3 CHEF gel mapping blot

Detailed procedures for direct mapping of genes to separated yeast chromosomes, and appropriate strains to use, have been worked out by several groups (38–41); the details of that approach are too lengthy to be included here.

3.2.4 Preparation of total yeast RNA

Protocol 13 describes the preparation of total yeast RNA using glass beads (42).

Protocol 13. Yeast RNA preparation

Reagents

- REB: 0.5 M NaCl, 0.2 M Tris-HCl, pH 7.5, 0.01 M EDTA, 1 per cent SDS.
- Equilibration buffer: REB, but without SDS.
- PCIA (24:24:1): Prepare fresh. Mix equal volumes of phenol and chloroform–isoamyl alcohol. Shake with an equal volume of equilibration buffer, and allow the phases to separate. Centrifuge briefly.
- 3 M NaOAc, pH 6.0: 21.5 g sodium acetate in 40 ml double-distilled water; adjust the pH with glacial acetic acid, then make the volume to 50 ml.

Procedure

1. Harvest 1–2×10^8 cells (more haploids than diploids are required, because they are smaller).
2. Resuspend the pellet in 200 ml REB. Transfer to an Eppendorf tube and freeze on dry ice.) The cells can be frozen at -70 °C for several weeks. Freezing seems to improve the efficiency of smashing, so freeze the samples briefly, even if completing the isolation right away.
3. Add about 0.3 g glass beads and 200 ml PCIA to each sample.
4. Thaw the pellets if they are still frozen. Vortex at top speed for 2.5 min.
5. Add 100 ml each of REB and PCIA. Vortex 1 min and centrifuge for 2 min in a microfuge.
6. Transfer the aqueous phase to a fresh tube, and re-extract with 200 ml PCIA. Centrifuge and transfer the aqueous phase to a fresh tube.

214

7. Add 2 volumes of 100 per cent ethanol; vortex. Freeze at least 10 min on dry ice, or store samples at this point. If you have more samples than you can vortex at once, carry each group of samples through to this point before thawing the next set of samples.

8. Centrifuge for 10 min; discard the supernatant.

9. To remove DNA, wash the pellet twice with 100 ml 3 M NaOAc, pH 6.0, at room temperature. Each time, use a pipette tip to grind the pellet, and pipette up and down several times to homogenize. Centrifuge for 10 min at 4 °C after each wash, and remove supernatant with a pipette.

10. Wash the pellet with cold 70 per cent ethanol. Vortex vigorously. Centrifuge for 5 min at 4 °C; remove the supernatant and dry the pellet.

11. Dissolve the pellet in 25 ml double distilled water. To quantitate, read A_{260} on 2 ml in 0.5 ml double-distilled water. Concentration in mg/l = $A_{260} \times 10$, for this dilution. 1 A_{260} = 40 mg/ml.

12. Calculate the volume of additional double-distilled water needed to adjust concentrations to 1 mg/ml. For gel electrophoresis load 1 to 4 µg in 3.5 mm lanes. Overloading will cause smearing of tubulin signal, because all three tubulin mRNAs comigrate with the 1.6 kb ribosomal RNA, which will smear the tubulin bands in heavily loaded lanes. The sizes of the ribosomal bands are 3.36 kb (25S) and 1.65 kb (18S).

3.2.5 Preparation of poly-A$^+$-RNA and Northern blotting

Prepare poly-A$^+$-RNA by the method of Maniatis (29), starting with the total RNA prepared as in *Protocol 13* above (43).

3.3 Preparation of yeast detergent-extracted cytoskeletons

Yeast cells, like other eukaryotic cells, can be selectively extracted to remove most soluble proteins but retain cytoskeletal structures, including microtubules, for subsequent morphological and biochemical analyses (44). *Protocol 14* describes the extraction procedure.

Protocol 14. Preparation of yeast cytoskeleton

Reagents

- PM2G: 15.1 g Pipes, 0.38 g EGTA, 0.12 g Mg_2SO_4, 92.1 g glycerol, adjust pH to 6.9 using 10 M NaOH. The solution will clear as the pH increases. Add water to 500 ml final volume.
- PM4G: PM2G with twice as much glycerol.
- NP40: Nonidet P-40 (BDH), 10 per cent solution in water. Store at −20 °C.

Protocol 14. *Continued*

Procedures

1. Harvest 1.5×10^8 cells and resuspend in 1.5 ml PM2G.
2. Add 3.75 ml β-mercaptoethanol and 3.75 ml 3 µg/ml Zymolase.
3. Incubate at 37 °C for 20–30 min.
4. During incubation, split into three tubes (Falcon 15 ml conical tubes), of 5×10^7 cells each for further analysis.
   - For *total protein analysis* (tube 1), go to step 5.
   - For *immunofluorescence analysis* (tube 2), and *protein fraction analysis* (tube 3), go to step 6.

Tube 1: Total protein analysis

5. Centrifuge and resuspend in PM2G with protease inhibitors. Smash, boil, and freeze using the procedure described in *Protocol 8.*

Tubes 2 and 3: Immunofluorescence and protein fraction preparation

6. After digestion (stage 3), centrifuge for 1.5 min at 800 rev/min (IEC centrifuge, small rotor).
7. Discard supernatants and resuspend pellets in 375 µl PM2G plus protease inhibitors at room temperature.
8. Add 19 µl 10 per cent NP40 to the side of each tube. Gently mix all samples at once by inverting several times. Centrifuge for 1.5 min at 800 rev/min (IEC centrifuge, small rotor).
   - For samples to be *fixed for immunofluorescence*, go to stage 9.
   - To *prepare protein fractions* go to stage 12.

Tube 2: Fixing for immunofluorescence

9. Remove supernatant; gently add 375 µl of 3.7 per cent formaldehyde.
10. Allow fixative to sit on pellet while you process protein fractions, then *gently* resuspend cells.
11. After 1 h, wash several times with 0.1 M potassium phosphate, pH 6.5.

Tube 3: Preparation of protein fractions

12. Take the upper half of the supernatant (197 µl) and add to 66 µl GSD. Boil 5 min in PM2G, freeze aliquots.
13. Discard the remaining supernatant. Add glass beads to pellet and resuspend in 300 µl PM2G plus protease inhibitors plus 100 µl GSD. Vortex 30 s, boil 5 min, freeze aliquots.

Notes

- 525 μl supernatant = 400 μl pellets = 5×10^7 cells.
- Using PM4G instead of PM2G seems to increase microtubule stability, but requires 3.5 min instead of 1.5 min centrifugations. Use PM2G for Zymolase digest, PM4G thereafter. PM2G works well on wild-type strains, but microtubules in the TUB2 hemizygotes were less stable to extraction.
- To avoid losing protein following extraction, boil the supernatant immediately, and vortex pellets only 30 s with glass beads. The total protein from tube 1 is necessary to assure quantitative recovery.
- Extracted cells can be very sensitive to shear forces when mounted on slides. Stain them on cover slips, and mount on gelvatol and 15 mg/ml DABCO (Aldrich Chemical Co.).

3.4 Immunofluorescence

The procedure described in *Protocol 15* is modified from ref. 45.

Protocol 15. Immunofluorescence

Reagents

- 1.2 M sorbitol in 0.1 M potassium phosphate, pH 6.5: 11 g sorbitol, 5 ml of 1 M potassium phosphate, pH 6.5, and to 50 ml final volume. Store at 4 °C (for a few days) or frozen at −20 °C (for longer periods). The buffer may be repeatedly thawed and refrozen.
- Permeabilization mix: 900 μl of 1.2 M sorbitol in 0.1 M KPi; 5.0 μl of β-mercaptoethanol, 7.5 μl of 3 mg/ml Zymolyase in SCE and 50 μl of Glusulase. Make up just prior to permeabilization reaction.
- Polylysine: 1 mg/ml polylysine; store in aliquots at −20 °C.

Procedures

Fixation and permeabilization

1. To a culture of cells in logarithmic growth add formaldehyde to a final concentration of 3.7 per cent (e.g. 1 ml 37 per cent formaldehyde stock plus 9 ml culture). Mix gently; leave at room temperature for 1 h.

2. Harvest by centrifugation; resuspend in approximately 1 ml 0.1 M potassium phosphate, pH 6.5, containing 1.2 M sorbitol. Transfer to an Eppendorf tube. Spin for a few seconds, remove the supernatant, and resuspend in same buffer. The fixed cells can be stored at 4 °C, or centrifuged immediately and resuspended in permeabilization mix.

3. Make up enough permeabilization mix (see above) for all samples, plus a

Protocol 15. *Continued*

bit extra, to allow for pipette inaccuracy. Add mix and incubate for 1 h at 37 °C. The volume for the permeabilization should be approximately 1/100 culture volume (e.g. 100–200 μl digest for 9 ml culture, depending on cell density).

4. Wash the permeabilized cells twice with 0.1 M potassium phosphate, pH 6.5 (without sorbitol). Store at 4 °C.

Immunofluorescence

1. Prepare a humidified box by placing a piece of moistened filter paper in the bottom of a plastic box; keep the slide (an eight-well multi-test slide (Flow Laboratories) or glass cover slip) in this box throughout the staining procedure. Apply 20 μl of 1 mg/ml polylysine to each well or 50 μl to cover slip; wait 5 min.

2. Wash by removing the drop with very gentle suction from an aspirator and immediately adding a drop of distilled water before proceeding to the next well. Wash each well three times. Wash cover slips by dipping three times in a beaker of water.

3. Apply 10 μl of cell suspension per well, or about 50 μl per cover slip. The suspension should be very faintly turbid. Try a range of concentrations at first. If the cells are too dense, they will be lost more readily in subsequent steps. After applying the cells, wait 10–15 min, then wash three times with phosphate-buffered saline (PBS). The suction of the aspirator should be as gentle as possible. There will be some buffer left behind after each aspiration.

4. Prepare the antisera; use PBS with 0.1 per cent BSA for all dilutions.

5. Apply 10–20 μl of antibody to each well. The larger volume seems to give more uniform staining. Use 50 μl on a cover slip; pipette up and down once or twice to mix with residual buffer. Incubate at room temperature for 1 h. Longer incubations will give increased background. Wash three times with PBS.

6. Apply the second antibody. Incubate in the dark at room temperature for 1 h. Wash five times with PBS.

7. Stain nuclei with DAPI (Boehringer-Mannheim), which is kept as a 1 mg/ml stock in water, stored at 4 °C. Use 15 ml of this stock in 40 ml water (the volume of a Coplin jar). At the end of the second incubation, after the washing procedure, dip the slide or cover slip sequentially into a beaker of water, the DAPI solution for 5 s, and another beaker of water.

8. For mounting the cover slip, use a solution of phenylenediamine in glycerol. This slows the bleaching of the fluorescent signal that results from illumination. Apply a generous amount of the mounting solution to

the slide, place the cover slip on top, and press between several layers of paper towels for 1/2 h to overnight. After pressing, seal the edges of the cover slip with nail polish, and gently clean the slide, if necessary. Keep the slide in an opaque box at all times to protect from bleaching. Store at 4 °C.

For viewing, we use a Zeiss microscope equipped for epifluorescence, with a Zeiss 487710 filter for fluorescein, or a Zeiss 487715 filter for rhodamine. These give the greatest separation between the two signals, should you want to do double-label immunofluorescence.

3.5 Electron microscopy

This complex topic has recently been reviewed (46). Specialized methods for analysing cytoskeletons are discussed in Chapter 2.

3.6 Anti-peptide antibodies

Raising antibodies against yeast proteins can be complicated by the high levels of anti-yeast protein antibodies present in pre-immune bleeds of most rabbits we have examined. Also, purifying minor yeast proteins in useful quantities can be difficult. Antibodies of exceptionally high titre, low background, and high specificity have been generated against the three tubulins of *S. cerevisiae* using peptides derived from their cDNA sequences. General procedures for raising such antibodies have been discussed at length by others (47). We have most success with peptides at least 10 residues long, hydrophilic, and preferably at the C- or N- terminus. We couple the peptides to a carrier protein, keyhole limpet hemocyanin (KLH), for injection into rabbits (see Chapter 4). Pre-immune sera should be obtained from each rabbit and tested by immunoblot analysis against total yeast proteins to eliminate those with high-titre anti-yeast antibodies, especially those blotting proteins in the size range of interest (a frequent occurrence). We typically use three rabbits per antigen.

References

1. Huffaker, T. C., Hoyt, M. A., and Botstein, D. (1987). *Ann. Rev. Genet.*, **21**, 259.
2. Drubin, D. (1989). *Cell Motil. Cytoskeleton*, **14**, 42.
3. Katz, W. S. and Solomon, F. (1989). *Cell Motil. Cytoskeleton*, **14**, 50.
4. Stearns, T. (1990). *Cell Motil Cytoskeleton*, **15**, 1.
5. Sherman, F., Fink, G., and Hicks, J. (1981). *Methods in yeast genetics*. Cold Spring Harbor Laboratory Press, New York.

6. Bruschi, C. V., Comer, A. R., and Howe, G. A. (1987). *Yeast*, **3**, 131.
7. Ito, H., Fukada, Y., Murata, K., and Kimura, A. (1983). *J. Bacteriol.*, **153**, 163.
8. Mortimer, R. K. and Hawthorne, D. C. (1975). *Meth. Cell Biol.* (*Yeast Cells*), **11**, 221.
9. Esposito, R. E. and Klapholz, S. (1981). In *Molecular biology of the yeast saccharomyces*. Strathern, J. N., Jones, E. W., and Broach, J. R. (eds.). CSH monograph, Cold Spring Harbor, New York.
10. Thomas, J. H. (1984). Genes Controlling the Mitotic Spindle and Chromosome Segregation in Yeast. Thesis, MIT.
11. Neff, N. F., Thomas, J. H., Grisafi, P., and Botstein, D. (1983). *Cell*, **33**, 211.
12. Kilmartin, J. (1981). *Biochemistry*, **20**, 3629.
13. Thomas, J. H., Neff, N. F., and Botstein, D. (1985). *Genetics*, **112**, 715.
14. Shortle, D., Novick, P., and Botstein, D. (1984). *Proc. Natl. Acad. Sci. USA.*, **79**, 1588.
15. Rose, M. D. and Fink, G. R. (1987). *Cell*, **48**, 1047.
16. Huffaker, T. C., Thomas, J. H., and Botstein, D. (1988). *J. Cell Biol.*, **106**, 1997.
17. Schatz, P. J., Solomon, F., and Botstein, D. (1988). *Genetics*, **120**, 681.
18. Katz, W. S. and Solomon, F. (1988). *Mol. Cell. Biol.* **8**, 2730.
19. Boeke, J. D., Trueheart, J., Natsoulis, G., and Fink, G. R., (1987). *Meth. Enzymol.*, **154**, 164.
20. Jarvik, J. and Botstein, D. (1975). *Proc. Natl. Acad. Sci. USA*, **72**, 2738.
21. Novick, P., Osmond, B. C., and Botstein, D. (1989). *Genetics*, **121**, 659.
22. Nasmyth, K. A. and Tatchell, K. (1980). *Cell*, **19**, 753.
23. Carlson, M. and Botstein, D. (1982). *Cell*, **28**, 145.
24. Rose, M. D. and Fink, G. R. (1987). *Cell*, **48**, 1047.
25. Hayles, J., Beach, D. H., Durkacz, B., and Nurse, P. M. (1986). *Mol. Gen. Genet.*, **202**, 291.
26. Hadwiger, J. A., Wittenberg, C., Richardson, M. E., De Barro Lopes, M., and Reed, S. I. (1988). *Proc. Natl. Acad. Sci. USA*, **86**, 6255.
27. Stearns, T. and Botstein, D. (1988). *Genetics*, **119**, 249.
28. Towbin, H., Staehelin, T., and Gordon, J. (1979). *Proc. Natl. Acad. Sci. USA*, **76**, 4350.
29. Maniatis, T., Fritsch, E. F., and Sambrook, J. (1989). *Molecular cloning: a laboratory manual*, 2nd edn. Cold Spring Harbor Laboratory, New York.
30. Glass II, W. F., Briggs, R. C., and Hnilica, L. S. (1981). *Science*, **211**, 70.
31. Glass II, W. F., Briggs, R. C., and Hnilica, L. S. (1981). *Anal. Biochem.*, **115**, 219.
32. Karcher, D., Lowenthal, A., Thormar, H., and Noppe, M. (1981). *J. Immunol. Meth.*, **43**, 175.
33. Holm, C., Meeks-Wagner, D. W., Fangman, W. L., and Botstein, D. (1986). *Gene*, **42**, 168.
34. Southern, E. M. (1975). *J. Mol. Biol.*, **98**, 503.
35. Church, G. M. and Gilbert, W. (1984). *Proc. Natl Acad. Sci. USA*, **81**, 1991.
36. Reed, K. C. and Mann, D. A. (1985). *Nucleic Acids Res.*, **13**, 7207.
37. Reed, K. C. (1985). *Bio-Rad Technical bulletin*, **1233**, 1.
38. Vollrath, D., Davis, R. W., Connelly, C., and Hieter, P. (1988). *Proc. Natl. Acad. Sci. USA*, **85**, 6027.
39. Volrath, D. and Davis, R. (1987). *Nucleic Acids Res.*, **15**, 7865.

40. Schwartz, D. C. and Cantor, C. R. (1984). *Cell*, **34**, 67.
41. Ferris, S. and Zoller, P. (1989). *Mol. Biol. Reports*, **7**, 6.
42. Warner, J. R. (1982). *Metabolism and gene expression*, CSH monograph, p. 542. Cold Spring Harbor, New York.
43. Broach, J. R., Atkins, J. F., McGill, C., and Chow, L. (1979). *Cell*, **16**, 827.
44. Pillus, L. and Solomon, F. (1986). *Proc. Natl. Acad. Sci. USA*, **83**, 2468.
45. Adams, A. E. and Pringle, J. R. (1984). *J. Cell Biol.*, **98**, 934.
46. Wright, R. and Rein, J. (1989). *Meth. Cell Biol.*, **31**, 473.
47. Harlow, E. and Lane, D. (1988). *Antibodies: a laboratory manual*. Cold Spring Harbor Laboratory, New York.

10

Studying intermediate filaments

ALBERTO DOMINGO, ALFONSO J. SARRIA, ROBERT M. EVANS,
and MICHAEL W. KLYMKOWSKY

1. Introduction

Intermediate filaments (IFs) are of interest both from the simple biological perspective, as intriguing cellular and supracellular structures, and at the more applied level, as important markers of cell type and tumour diagnosis in development and surgical pathology (see refs. 1–6 for reviews). IFs are found in the nucleus and in the cytoplasm of eukaryotic cells. In the nucleus, IFs are a major component of the nuclear lamina, which appears to be a defining (i.e., ubiquitous) characteristic of eukaryotic cells. Cytoplasmic IFs are found in a wide range of metazoans (7), and can compose a substantial percentage of the total cellular dry mass. In addition to the *bona fide* IF proteins, there is also clear evidence for proteins that are related to various extents to IF subunit proteins. For example the tektins, which are microtubule-associated proteins, cross-react immunologically with the IFs of vertebrates and may share some structural features (8). Nevertheless, the tektins are not IFs in any meaningful sense, and their cross-reaction with IF proteins underlines the care that must be taken in defining unknown proteins as IF proteins.

After the identification of a new putative IF protein or its coding gene there are three well-defined, yet often interdependent, levels in the study of IF structure and function. The most readily apparent activity of IF subunit proteins (IFsps) is to polymerize into IFs. Understanding how IFsps assemble into IFs and how IFs interact with other cellular structures involves methodologies common to all studies of protein–protein interactions. The close correlation between IFsp gene expression and cell type makes IFsps important markers of cellular differentiation. In the case of malignant transformation, IFsps can also be used to trace tumour origin. These studies involve immunocytochemistry and *in situ* hybridization, together with the analysis of specific *cis*- and *trans*-acting factors that control IFsp synthesis. Finally, understanding the role of IFs in the physiology of the cell, tissue, organ, and the complete organism remains the greatest challenge to our understanding of IFs. Despite a great effort devoted to the solution of this question over the last ten years, the true functions of IFs remain unclear (4).

223

Our goal here is to provide the reader with a handy reference that should make studies of all three aspects of IFs more straightforward.

2. Defining IF subunit proteins

The term 'intermediate filaments' was first applied by Ishikawa *et al.* (9) to the morphologically distinct set of cytoplasmic filaments present in muscle cells. With diameters of roughly 10 nm, these filaments were intermediate between the thin (actin) and thick (myosin) filaments of the sarcomere. Later studies showed that these muscle IFs are related to the wool 'keratins', epithelial 'tonofilaments', neuronal 'neurofilaments', and the insoluble filamentous network found in many other cell types. More recently, the relationship between cytoplasmic IFs and the eukaryotic nuclear lamins was discovered (10, 11). New cytoplasmic IFsps have been described recently, e.g. α-internexin (12, 13), peripherin (14, 15), and nestin (16); presumably more await discovery (see 17, 18). In addition, there may well be other proteins that share primary and secondary structural similarity with IFsps, but which fail to form IFs (19). This raises an important practical question when dealing with a previously uncharacterized protein, namely, how is one to decide whether it is or is not a true IFsp. Despite their diversity, all IFsps share well-defined characteristics that enable us to answer this question unambiguously. These include a highly conserved primary and secondary structure, an ability to renature and form 10 nm filaments, and a characteristic structure and insolubility of these filaments.

All types of IFsps share a highly stereotyped secondary structure. This consists of a central α-helical 'rod' domain, flanked by non-helical amino-terminal 'head' and carboxy-terminal 'tail' domains. The central rod domain is subdivided into four α-helical stretches characterized by a repetitive motif of seven amino acids in which the first and fourth amino acids are hydrophobic (i.e. a 'heptad repeat'). These four regions fold into an extended α-helix in which the hydrophobic residues form a continuous helical stripe (20). This structure is stabilized in aqueous media by association into coiled-coil dimers. The stretches of heptad repeat are separated by three 'linker' regions that do not conform to the heptad repeat structure and often contain the α-helix-destabilizing amino acid residue proline. The length of the central rod domain (310 amino acids) and the position of the linkers is highly conserved among all of the cytoplasmic IFsps of vertebrates that have been analysed to date. In invertebrates (21, 22) and the nuclear lamins (10, 11), the central rod domain is 42 amino acids (6 heptads) longer, and the linker domains do not contain Pro residues. The head and tail domains that flank the rod domain display little recognizable structure, at least as discerned from currently available sequence analysis methods (3).

In all known IFsps, acidic amino acid residues constitute 20–25 per cent of the molecule. This high number of negative charges is generally not compensated by basic amino acid residues, and so most IFsps have acidic isoelectric points and a net negative charge at neutral pH. The only known exceptions to this rule occur among the cytokeratins, some of which have neutral or basic isoelectric points (23, 24). In addition to the common properties of the central rod domain of IFsps discussed above, the terminal domains also have characteristic sequence properties. For example, although the homopolymeric IFsps have acidic isoelectric points, the amino-terminal domains are characterized by sequences that are rich in hydroxyamino acids and arginine, with virtually no acidic residues (25).

Several post-translational modifications have been found to occur in essentially all IFsp types. Most of these modifications are localized to the non-helical terminal domains, due either to the greater accessibility of these domains when the IFsp is assembled into an IF, or to the presence of the specific target sequences. Modifications observed in IFsps include the acetylation of the N-terminal residue (1, 26, 27), phosphorylation (see below), ubiquination (28, 29), and interchain cross-linking by trans-glutaminases (30, see 1). The nuclear lamin proteins A and B are isoprenylated (31–33), and the glycosylation of cytokeratins (34) has been reported. The physiological significance of most of these modifications is not yet well understood.

Inter- and intramolecular oxidation of cysteines into disulphide bonds are infrequent in IFsps *in vivo*. Only in keratinocytes have disulphide bond cross-linked cytokeratins been detected (35, 24, 36). Moreover, a number of IFsps have no Cys residues at all. *In vitro* disulphide bridges may form artifactually during protein purification. For this reason reducing agents should be included during the manipulation of IFsp samples (see below).

All IFsps are phosphorylated to some extent in the interphase cell. This modification can substantially affect the isoelectric point and, in the case of neurofilaments, the antigenicity and electrophoretic mobility of the proteins (37, 38). Phosphorylation has been implicated in structural reorganizations of nuclear and cytoplasmic IFs (see 39 and references therein, 40–48). In *in vitro* systems, a number of protein kinases have been shown to phosphorylate IFsps (49–54). Modification of IFsps *in vitro* can alter the assembly properties of the subunit proteins, but there are significant differences in the results obtained with different types of IFsps. *In vitro* phosphorylation induces IF disassembly in the case of vimentin and desmin (51–53, 55–58), and conformational changes (59), but not disassembly (60), in the case of cytokeratins. In the neurofilament NF-M and NF-H proteins, dephosphoryl-ation is reported to either have no effect (38) or to dissociate the IFsps from assembled filaments (37). Phosphorylated NF-L is incompetent to assemble; however, phosphorylation of NF-L filaments stabilizes the polymer, reducing the rates of disassembly and subunit exchange (61). Some of these observed differences in the effects of phosphorylation on the assembly state of IFsps probably reflect differences in the location of phosphorylation sites. In addition, the relative amount of IFsps that is actually modified with phosphate at a given site is an important parameter (51) that is often not determined and may not be comparable in different studies.

Studying IFsp phosphorylation *in vitro* allows some experimental approaches that are difficult to carry out in intact cells. Carefully defined conditions permit the action of individual protein kinase activities to be identified and differentiated. Because the specific activity of the phosphorylated protein that can be produced *in vitro* is many times greater than can be achieved in intact cells, subsequent analysis of phosphorylation sites is greatly simplified. Although *in vitro* phosphorylation experiments are often useful, some caution should be exercised in directly extrapolating results from these studies to living cells. In contrast to the effects of phosphorylation of IFsps *in vitro*, the intracellular injection of unregulated kinases induces the 'collapse' of the system, but not detectable disassembly of IFs (62). It has also been reported that the soluble pool of vimentin in cells does not appear to be more highly phosphorylated than the assembled IFsps (63, 64, 65). This leaves unresolved the actual physiological significance of *in vitro* phosphorylation.

2.1 Proteolysis of IFsps

Calcium-activated, neutral thiol proteinases, i.e. calpains, are believed to play a prominent role in the intracellular turnover of IFsps. Nelson and Traub (66) first described a calpain that specifically degrades vimentin and desmin *in vitro* (see also 67, 68). This proteinase preferentially degrades the non-α-helical regions of vimentin (67, 69). *In situ* neuro- and glial filaments (see 70, 71), as well as vimentin (72), have been reported to be degraded by calcium-activated proteinases. Because of the widespread occurrence of IFsp-specific proteinases, care must be taken to avoid artifactual proteolysis during protein purification. Typically, chelation of calcium ion and the inclusion of various proteinase inhibitors, such as $0.5-1$ µg/ml leupeptin (72), will suffice. If these prove inadequate, then samples can be frozen rapidly, in either liquid nitrogen or acetone–dry ice, and then homogenized to minimize proteolytic activity.

2.2 General biochemical characteristics of IF proteins

The biochemical characteristics of IFs and IFsps are homogeneous and peculiar enough to form a first step in the definition of a protein as an IFsp. IFs are among the most insoluble structures of the cytoskeleton and their extreme insolubility is generally exploited in their purification, since it permits other, more soluble cellular proteins to be washed away. A second biochemical property of IFsps is their ability to form IFs following renaturation. After urea, guanidine, or SDS has been used to solubilize the IFs (*Table 1*), the removal of the denaturant generally leads to the reassembly of IFsps into IFs (see below). Under these conditions, the physical form of the

Table 1. Denaturing reagents used for IFsps

Reagent	Characteristics
urea	final concentration 6–9.5 M
	compatible with SDS; does not affect SDS-PAGE
	suitable for gel filtration[a] or ionic exchange chromatography and fractionation with ammonium sulphate (or any other salt)
	IFsps dissolved in urea precipitate at $\geq$ 40 per cent saturated ammonium sulphate[b]
	proteinases are inactive in 8 M urea
	hydrolysis products react with amino groups, affecting the pI[c]
guanidine	final concentration 4–7 M
	incompatible with SDS; strongly affects SDS-PAGE
	suitable for gel filtration[a] but not ionic exchange chromatography or salt fractionation
	proteinases are inactive in 6 M guanidine
	proteins dissolved in guanidine can be precipiated with 9 volumes of ethanol/ether, 1:1, at −70 °C for 1 h
SDS	final concentration 5–6 per cent (w/v)
	does not denature amphiphatic α-helix, but dissociates IF oligomers; many proteinases remain active in SDS
	dialyses slowly (forms large micelles)
	suitable for gel filtration[a] but not ion exchange chromatography or salt fractionation

[a] The effective size of unfolded proteins is much larger than the native conformation, and also varies with the nature of the denaturant (see refs. 74, 75).

[b] Amount corresponding to this percentage of saturation in water at 20 °C (using standard tables). Saturation in 8 M urea corresponds to approximately 75 per cent saturation in water.

[c] Urea hydrolyses spontaneously in water, reaching equilibrium in a few hours at room temperature.

IFsps varies depending on the concentration of denaturant. Initially in a monomeric and unfolded state, as the concentration of denaturant decreases, IFsps first form dimers, then tetramers, then higher-order oligomers, and finally IFs (73).

While insoluble in high ionic strength media, IFs do show partial and sometimes complete solubility in very low ionic strength buffers. Homopolymeric IFs are in general more soluble than the heteropolymeric IFs (cytokeratins) in low-salt media. Velocity sedimentation centrifugation on sucrose gradients (5–30 per cent or 10–40 per cent w/v sucrose gradient at 36 000 rev/min for 18 h in a Beckman SW40 rotor at 10 °C) can be used to define the size of soluble IF oligomers, particularly as isolated from cultured cells (64, 76). Tetrameric (77, 64) and higher-order oligomers (183) of IFsps have been described. The reported sedimentation coefficients for IF tetramers range between approximately 5 S and 6.7 S (77, 64). Isopycnic centrifugation has been used to show associations of IFsps with lipids (78–80) and nucleic acids (81–86). Protocols suitable for studying lipid association with IFsps are given in Chapter 6 of this volume.

2.3 Isolation of the IF-rich cytoskeletal fraction

In cultured cells, it is relatively simple to isolate the IF-rich 'cytoskeletal' fraction. The method we typically use involves washing cells grown in flasks or Petri dishes with divalent cation-free phosphate-buffered saline (PBS) containing 5 mM EGTA and then extracting the cells with PBS containing 5 mM EGTA and 5 per cent NP40. In some cases the detergent extraction will release the cells from the substratum; if so, the insoluble components can be recovered by centrifugation at 1000 rev/min for 10–15 min in a clinical centrifuge. If required, the detergent extraction can be followed with a high-salt wash (1.5 M KCl), although in most cases this is unnecessary. The resulting insoluble material can be collected by low-speed centrifugation and is depleted in cytosolic components and the cytoskeletal components tubulin, actin, and myosin. However, a number of non-IF insoluble components remain, such as remnants of nuclei and the extracellular matrix. If desired, high-molecular-weight DNA and associated proteins, as well as residual filamentous actin, can be removed by briefly treating the detergent-insoluble material with 0.5 mg/ml DNAse I and 10 mM $MgCl_2$.

The major modification required to prepare IF-rich fractions of tissues involves the need to inhibit the generally higher level of proteolytic activity. The specimen is homogenized in a physiological salt buffer in the presence of a non-ionic detergent, such as Triton X-100 or NP40, to solubilize cytosolic and membrane-associated components. The insoluble fraction is collected by centrifugation and resuspended in high-salt buffer to solubilize actin and myosin. The insoluble residue is composed primarily of IFs. In tissues that contain a large amount of connective tissue, such as muscle, the insoluble residue will also be enriched for these components. Further purification (see below) requires solubilization with a strong denaturant, such as urea, guanidine or SDS (*Table 1*). The expression of cDNAs in bacteria is rapidly becoming a major source of IFsps for *in vitro* studies (87–89). Once purified, bacterially-synthesized IFsps appear to polymerize quite effectively (see 89). A procedure for isolating IFsp inclusion bodies from *E. coli* is given in *Protocol 1*.

Protocol 1. General purification protocol of IFsps expressed in bacteria

Growing cells and inducing protein expression (assuming that the expression vector is selectable by ampicillin resistance and inducible by IPTG)

● LB media (per litre): 10 g bactotryptone, 5 g bactoyeast extract, 10 g NaCl.

● Ampicillin stock solution: 100 mg/ml in water, 0.22μ filtered. Store at −20 °C.

● IPTG stock solution: 100 μM in water, 0.22μ filtered. Store at −20 °C.

1. Inoculate 5 ml LB + 100 µg/ml ampicillin with a single colony from an agar culture plate. Incubate overnight at 37 °C with shaking. Autoclave a 2 l flask with 500 ml LB, and let it stand overnight at 37 °C.

2. Add ampicillin to the prewarmed 500 ml LB to a final concentration of 100 µg/ml. Inoculate with 1 ml of the 5 ml overnight culture. Incubate at 37 °C with strong shaking until the culture reaches 0.6 OD_{550nm}.

3. Induce protein expression by addition of IPTG to a final concentration of 0.4 mM. Incubate for 2 h at 37 °C with shaking.

Isolation of inclusion bodies (adapted from ref. 90.)

• Lysis buffer: 25 per cent sucrose (w/v), 50 mM Tris-HCl, 1 mM EDTA, pH 8.0.

• DNase buffer, 100×: 1 M $MgCl_2$, 100 mM $MnCl_2$.

• Detergent buffer: 0.2 M NaCl, 1 per cent deoxycholic acid (w/v), 1 per cent NP40 (v/v), 20 mM Tris-HCl, 10 mM NaF, 2 mM EDTA, pH 7.5

• Triton buffer: 0.5 per cent Triton X-100, 1 mM EDTA, 0.1 M NaCl, 10 mM Tris-HCl, pH 7.5.

1. Harvest cells by centrifugation at 5000 × g for 10 min at 4 °C. Decant the supernatant; disaggregate the pellet by vortexing. Resuspend the cells in 100 ml ice-cold lysis buffer. Aliquot into four 50 ml tubes and pellet the cells by centrifugation at 5000 × g for 10 min at 4 °C. Disaggregate the pellet by vortexing. The following steps are for one of the four tubes.

2. Freeze in dry ice for 15 min or longer. Do not omit this step. Cells can be stored at −70 °C for weeks.

3. Thaw rapidly by addition of 8 ml of lysis buffer and vortexing. Resuspend the cells completely by pipetting if necessary. Add 2 ml of 20 mg/ml lysozyme in lysis buffer. Let stand on ice for 30 min.

4. Add 100 µl of 100× DNase buffer and 100 µl of 1 mg/ml DNase I. Incubate at 37 °C for 30 min.

5. Add 20 ml of detergent buffer and vortex. Centrifuge the lysate at 5000 × g for 10 min; discard the supernatant.

6. Resuspend the pellet in 5 ml of lysis buffer. Add 50 µl DNase buffer and 50 µl of 1 mg/ml DNase I solution. Incubate at 37 °C for 30 min.

7. Add 10 ml of detergent buffer and vortex. Centrifuge at 5000 × g for 10 min; discard the supernatant.

8. Resuspend the pellet in 5 ml of Triton buffer. Centrifuge at 5000 × g for 10 min; discard the supernatant. Repeat this step twice using 1 ml of buffer and centrifuge in a microcentrifuge.

9. Solubilize the inclusion body pellet in 9 M urea, 50 mM Tris-HCl, 60 mM 2-mercapthoethanol, pH 8.0.

The ability of IFsps to renature and polymerize *in vitro* provides an effective purification technique under some circumstances. This method consists of successive cycles of solubilization by denaturation and subsequent renaturation and polymerization by dialysis, recovering the insoluble, IF-rich, fraction by centrifugation. Proteins that remain in solution after the dialysis are readily eliminated by this procedure. The efficiency of the technique is limited by several factors. First, the great majority of protein species do not renature as readily as IFsps and, as a general rule, denatured proteins have limited solubility. Thus, they are recovered in the insoluble fraction after centrifugation. Second, contaminants that are able to bind to IFsps or IFs, whether physiologically significant or artifactual, will copurify by this method. Finally, the ability of IFsps to renature and polymerize depends very much on their α-helical and coiled-coil nature. Both features are also present in other cytoskeletal components, in particular in myosin and its naturally occurring peptide fragments. Not only the biochemical properties, but also the ultrastructural morphology of myosin and myosin fragment polymers, is remarkably similar to IFs (Domingo and Marco, unpublished observations).

When considering the utility of polymerization-depolymerization as a method for purifying IFsps, it is important to consider the source of the IFsps. In particular, the inclusion body fractions isolated from bacteria (see *Protocol 1*) are particularly rich in proteolytic enzymes. These enzymes, while not active in concentrated urea or guanidine solutions, often recover their hydrolytic activity during dialysis. In our experience, the immediate application of repolymerization protocols to crude urea- or guanidine-solubilized fractions is not recommended.

2.4 Chromatographic purification of IFsps

To complete the purification of an IFsp from either an insoluble cytoskeletal fraction or a bacterial inclusion body preparation requires one or more chromatographic steps. The IFsps need to be dissolved in a strong denaturant. The behaviour of IF proteins in all types of chromatographic separations depends very much on the nature and concentration of the denaturing reagents in the running buffer. The study of cytokeratin oligomers is further complicated by the fact that cytokeratins form extraordinarily stable complexes. Purified type I and type II cytokeratins alone exist as monomers in 6 M urea/reducing agents. Under these same conditions, a mixed solution of type I and II cytokeratins forms tetramers (89). Even in 9 M urea, mixtures of type I and type II cytokeratins still form dimers and tetramers. Oligomers and free monomers can be separated by anion exchange Mono-Q FPLC columns or by gel filtration on Superose 6 FPLC columns (see below).

2.4.1 Ion exchange chromatography

Anion exchange chromatography is generally a good initial choice, given the acidic nature of most IFsps, and can be used for neutral-basic IFsps with alkaline pH buffers. DEAE-cellulose resins for conventional chromatography and Mono-Q columns (Pharmacia-LKB) for fast protein liquid chromatography (FPLC) have been used successfully with good results. In the case of bacterially synthesized cytokeratins, we routinely use a Mono-Q FPLC column with 9 M urea, 50 mM Tris-HCl, 2 mM DTT, 0.3 mg/ml PMSF, pH 8.1, as the sample buffer (89). The column is then eluted using a linear gradient of 0 mM to 200 mM guanidine–HCl in the same buffer. Similar conditions work well with conventional DEAE-cellulose and DEAE-Sepharose chromatography. Other running conditions have been used with comparable results for the purification of other intact and modified IF proteins (91).

2.4.2 Single-stranded DNA (ssDNA) chromatography

The observation that vimentin binds to ssDNA *in vitro* led to the use of ss-DNA cellulose as an affinity matrix for the rapid purification of IFsps (92) (*Protocol 2*). Although it is not clear whether all IFsps will bind to ssDNA cellulose, ssDNA associations in solution with desmin (82), GFAP (1), neurofilament proteins (85) and nuclear lamins (93) have been reported. Single-stranded DNA-cellulose, usually calf thymus DNA, is commercially available from a number of suppliers. This procedure is particularly useful in the isolation of IF protein from dilute solutions. Although ssDNA chromatography can be an effective purification step in the removal of non-IF proteins, it reportedly does not separate vimentin from bound lipids (78).

Protocol 2. Purification of vimentin by ssDNA cellulose
chromatography (92)

1. Prepare a ssDNA column. Commercial ssDNA cellulose usually contains 3–6 mg calf thymus DNA/gm. 1 g ssDNA cellulose (approximately 4 ml bed volume) will bind up to 1 mg of vimentin. Hydrate the powder in 50 mM Tris-HCl, pH 7.2, 6 M urea, 1 mM EDTA, and 1 mM dithiothreitol (Tris-urea buffer). Pour into a column and wash extensively with Tris-urea buffer containing 0.4 M KCl until there is no detectable absorbance at 260 nm in the eluate. Equilibrate the column in the Tris-urea buffer without KCl.

2. Prepare an IF-rich cytoskeletal extract (section 2.3). If chromatographically pure vimentin is desired, anion exchange chromatography can be performed at this point.

3. Add solid urea to the extract to a final concentration of 6 M and resuspend the precipitate. Centrifuge at 10–15 000 × *g* for 5–10 min and recover the supernatant.

4. Apply the urea-solubilized material to the ssDNA cellulose column equilibrated in Tris-urea buffer containing 0.5 μg/ml leupeptin. Wash the column with at least 10 column volumes of the same buffer.

Protocol 2. *Continued*

5. Elute a 4 ml column with a 30–50 ml linear gradient of 0–300 mM KCl and collect fractions. The vimentin content of the fractions should be analysed by SDS-PAGE.

6. Pool vimentin-containing fractions and dialyse extensively to remove the urea.

2.4.3 Gel filtration chromatography

In gel filtration chromatography, protein in an unfolded state has an apparent effective size (Stokes radius) that is much larger than the same protein in its native conformation. The type of denaturing reagent used in the running buffer will also affect the apparent size of the protein, so gel filtration matrices must be chosen with a separation range that is larger than the actual size of the protein(s) of interest. The performance of the gel filtration chromatography under denaturing conditions is otherwise as good as, or even a little better than, in non-denaturing buffers (74, 75). This phenomenon is the only peculiarity of gel filtration applied to IFsps. Running buffers used for gel filtration chromatography of IFsps include 50 mM Tris–HCl, pH 7–8.8, 10 mM β-mercaptoethanol, with 4 M to 7 M guanidine–HCl or 6 M to 9 M urea. SDS (1 per cent) can be included in urea, but not in guanidine buffers.

Samples can be boiled in buffers containing guanidine or SDS, but heating in the presence of urea may produce extensive covalent modification of the proteins by carbamylation of the amino groups (see *Table 1*). Suitable gel filtration matrices for conventional chromatography are Sephacryl S-300 or Sepharose CL-4B, and Superose 12 or Superose 6 columns for FPLC. In standard gel filtration chromatography, the high density of concentrated urea or guanidine solutions should be considered. Rigid gel matrices with good pressure resistance should be selected, and care should be taken to avoid the application of high hydrostatic pressures that may collapse the column.

2.4.4 High-performance liquid chromatography

Reversed-phase high-performance liquid chromatography (HPLC) has been used for preparative purification of assembly-competent cytokeratin proteins from crude detergent-insoluble cytoskeletons on CN-silica columns (77, 64). Vimentin and desmin have also been purified using C4 bonded silica columns (42) (*Protocol 3*). The capacity of HPLC is limited primarily by the solubility of intact IF proteins. HPLC is rapid and highly reproducible, and because the mobile phase is volatile, it yields purified protein in a salt-free form. Reversed-phase HPLC is mainly appropriate for small-scale, semi-preparative applications and most useful for the separation of mixtures of IFsp peptide fragments, either for analysis or as a preparative step for composition or sequence analysis (42).

Protocol 3. Reversed-phase HPLC analysis and purification of IFsps (modification of Evans (42))

1. Prepare an IF-rich cytoskeletal extract (Section 2.3).

2. Resuspend the precipitate in a small volume of 7 M urea. Centrifuge at 12 000 $\times$ g for 5 min and recover the supernatant. Add 0.1 volume (v/v) 1 per cent aqueous trifluoroacetic acid (TFA) and centrifuge again to remove any precipitated material.

3. Apply to a 250 mm $\times$ 4.6 mm C4 bonded silica reversed-phase column equilibrated in aqueous 0.1 per cent TFA. Run the column at room temperature at a flow rate of 1.5 ml/min.

4. Elute retained material with a linear gradient of 0–50 per cent acetonitrile in 0.1 per cent TFA and collect 1 ml fractions. If the starting material is an NP-40 or Triton X-100 extract, the early-eluting, major A280-absorbing peak is detergent. The next major eluting peak in non-proteolysed samples is vimentin.

5. Pool peak fractions with significant A280 absorbance, dry under vacuum, and resuspend in a buffer of choice. Although the retention time of IFsps on reversed-phase HPLC is highly reproducible, the identity of proteins in individual fractions in initial HPLC runs should be verified by SDS-PAGE.

2.4.5 Preparative SDS-PAGE

Vimentin (94, 55) and lamins A, B and C (93) have been purified by preparative SDS polyacrylamide gel electrophoresis (SDS-PAGE) of Triton-insoluble cytoskeletons (*Protocol 4*). Under these conditions up to 300 μg of vimentin can be recovered from a single preparative gel. Because of the polymerization chemistry, polyacrylamide gels contain an oxidative environment. To prevent significant protein oxidation, gels should be prepared at least 24 h before use and 0.5 mM thioglycolic acid should be included in the cathode running buffer to provide a reducing agent that will run ahead of the protein during electrophoresis (95). This technique is a simple, inexpensive, and relatively rapid method of producing small amounts of highly purified protein, with a minimum of proteolysis. It is particularly useful in studies on protein phosphorylation, because the protein can be purified from samples containing significant quantities of unincorporated ^{32}P under conditions that limit the exposure of laboratory personnel to the isotope.

Protocol 4. Purification of IFsps by preparative SDS-PAGE (modification of Evans (94))

1. At least 24 h prior to use, pour a 7.5 per cent SDS-PAGE gel. Thin gels (< 1 mm) give the best resolution, while thicker gels (1–2 mm) have a greater loading capacity.

2. Prepare an IF-rich cytoskeletal extract (Section 2.3).

3. Resuspend the precipitate in a solution containing 6 per cent SDS, 0.14 M Tris-HCl, pH 6.8, 1 mM dithiothreitol (DTT), 20 per cent glycerol, and 0.001 per cent bromophenol blue.

4. Apply the sample across the gel and electrophorese with 0.5 mM thioglycolic acid added to the cathode (upper) running buffer.

5. Following electrophoresis, stain the gel briefly with a solution containing 20 per cent methanol, 3 per cent acetic acid, 0.1 per cent Coomassie Blue. At high concentrations (100–300 µg/14 cm wide × 1 mm thick gel) IFsps bands will be faintly visible in minutes. To obtain assembly-competent IFsps, do not expose the gel to acidic methanol for prolonged periods.

6. When the protein bands are visible, rinse the gel in 0.1 M ammonium bicarbonate, 5 mM DTT until the solution is no longer acidic. Using a razor blade, cut out the stained protein band of interest.

7. Elute the protein from the gel slice by shaking in a buffered solution containing 0.1 per cent SDS, 1 mM DTT. Use 2–3 changes of this solution and combine. Recover the protein and remove the SDS by precipitation in cold acetone. The protein can also be electroeluted with similar recoveries if an apparatus is available.

3. IF oligomers and filaments

IFsp oligomers are of interest as the building blocks of the IFs in the cell (see below). They are also useful for the direct study of the molecular basis of their remarkable interchain interactions, among the most stable thus far described in nature. In current models of IF assembly, two monomers interact via their rod domains, in parallel and in register, to form a coiled-coil dimer, in which homologous stretches of the two strands are very closely apposed. IFsps can be divided into two broad categories, based on their ability to function as homodimers. Vimentin, desmin, GFAP, NF-L are able to form functional homodimers and can co-dimerize efficiently with one another. In contrast, the cytokeratins do not form functional homodimers (96). Instead, a type I and a type II cytokeratin assemble into a heterodimer (96, 89). This establishes the strict 1:1 stoichiometry of type I and type II cytokeratins required for cytokeratin filament assembly. It appears that any number of type I and type II cytokeratins can be involved, as long as the overall ratio of type I to type II cytokeratins is 1:1.

Two dimers align side by side to form a tetramer, often considered a stable intermediate in IF assembly. The arrangement and orientation of the dimeric coiled coils within the tetramer is not known with certainty. Further lateral and end-to-end associations of tetramers leads to the formation of an IF. According to this model, the end domains of the IFsps are located outside the filament backbone, which consists of the α-helical rod domains (see 3, 73 for reviews). How the pattern of IFsp assembly *in vitro* relates to the assembly of IFs *in vivo* remains unclear. Neither monomeric nor dimeric IFsps have ever been recovered from living cells. Small amounts of tetrameric IFsps have been found (64, 76), but whether these are intermediates in the assembly of IFs has not been determined. The very real possibility exists that *in vivo* IF assembly is assisted by assembly factors or chaperonins, very much in the way proteins such as nucleoplasmin aid in the assembly of nucleosomes (97).

3.1 Do the head and tail domains play a role in IF polymerization?

Several proteolysis studies have shown that substantial amounts of head and tail peptides can be released from keratin IFs without affecting the morphologic integrity of the filaments (see 98, 99). Recently, Lu and Lane (100) systematically examined the requirements for cytokeratin filament assembly *in vivo* using artificially constructed mutant proteins. They found that the presence of both head and tail domains on at least one of the two cytokeratins was necessary and sufficient for IF formation. This suggests that head and tail domains play a role in the end-to-end association of tetramers, but that there is sufficient redundancy in the interaction to tolerate some 'headless' and 'tail-less' polypeptides (see also 101, 102). A similar conclusion can be drawn from studies of proteolytically truncated or genetically engineered headless or tail-less mutants of desmin (103, 104), the NF-L and NF-M proteins (105, 106), and vimentin (Cary and Klymkowsky, unpublished observations). The mutant proteins can coassemble with wild-type IFsps and forms IFs; on their own, however, headless or tail-less IFsps fail to form extensive IF systems (104, 107, Carey and Klymkowsky, unpublished observations).

3.2 Forming IFs *in vitro*

At first glance, there is a surprising diversity among the published protocols for *in vitro* IF polymerization. In fact, these protocols are really minor variations on the same theme, which is to bring the sample from denaturing conditions to a more or less physiological medium in a progressive fashion. Since the exact nature of the intermolecular interactions that take place in the polymer is unknown, the final conditions selected for the polymerization of any specific IFsp are necessarily empirical. The starting material usually consists of IFsps in a denaturing solution, e.g. 4–7 M guanidine–HCl or 6–9 M urea, often in an alkaline buffer (pH 8.0–8.5). In the case of homopolymeric IFsps, the starting preparation may be in the form of soluble oligomers in a buffer of low ionic strength. Both denaturing solutions and dialysis buffers should include β-mercaptoethanol, dithiothreitol, or dithioerythritol as reducing reagents to prevent the formation of disulphide crosslinks between the IFsps. When polymerization conditions for a new, putative IFsp are being determined, we suggest that ionic strength should be varied first, followed by pH, and finally the chemical nature of the salt and buffer. A relatively long dialysis time of 12 h or more is recommended, as well as low temperatures and the addition of proteinase inhibitors to avoid degradation if the sample is not completely pure. Finally, it should be remembered that phosphorylation, and perhaps other post-translational modifications, can affect the polymerization ability of IFsps. *In vitro* IF assembly requires only that the subunit protein be above a critical concentration. The exceptions are NF-M and NF-H, which polymerize efficiently only in the presence of NF-L or another homopolymeric IFsp (see ref. 1). IF assembly then proceeds spontaneously without requirement for divalent cations, nucleotides, or other exogenous factors (108–111).

Electron microscopy is used to evaluate polymerization conditions. The methods used are relatively straightforward, ranging from simple negative-stain techniques with uranyl acetate to high-resolution metal shadowing. Uranyl acetate negative-staining of IF spreads is a very easy and useful technique for many studies of IFs. The preparation of grids and the spreading and negative staining are illustrated in *Protocol 5*. A rough determination of the percentage of protein in insoluble form and protein that remains in solution may be used to quickly narrow the range of suitable polymerization conditions.

Protocol 5. Negative stain electron microscopy for IFs

1. Prepare a clean beaker filled with distilled water and remove dust from the surface using a paper tissue.

2. Clean a glass microscope slide free of dust and immerse it in a 6 per cent solution of formvar (in chloroform). Take the slide out of the solution with a single and continuous movement to avoid horizontal stripes of different thickness in the film.

3. Immediately apply the edge of the slide to filter paper to absorb completely the excess of solution. Let the slide dry.

4. Using a razor blade, scrape the long edges of the slide and cut the surface of the film a couple of mm above the short side. Both faces of the slide can be used.

5. Breathe once on the film-coated surfaces (optional). Introduce the slide into the water-filled beaker slowly until the cut in the film parallel to the short side of the slide delineates sharply the meniscus of the liquid. Proceed very slowly while the surface tension begins to separate and peel off the film from the glass, then continue until the whole film floats freely. Move the slide horizontally as far away from the film as possible and then take it out of the water.

6. Judge the quality of the film by the homogeneous diffraction of light. Arrange the grids (200–300 mesh) on the film. Avoid breaking the film; repositioning the grids is impossible after they contact the surface.

7. Take the film and grids out of the beaker, using a piece of Whatman 3MM paper or parafilm. Lift with a quick continuous movement and let dry completely in a Petri dish. Carbon-coat the grids using a vacuum evaporator and glow-discharge the surface to enhance the adherence of proteins.

8. Cut the film around a grid with the tip of the forceps and lift it. Holding the grid with the forceps, deposit 5 µl of suspension of IFs on the coated surface and wait a couple of minutes. Rinse the grid with 100–200 µl of 1 per cent uranyl acetate in water. Remove the excess liquid by briefly touching the edge of the grid to a piece of filter paper. Place grids in a storage box and let them dry completely before examination.

3.3 Studying IF dynamics *in vitro*

For some time, IF were considered to be relatively static structures. Recent *in vitro* (61) and *in vivo* (112, 113) studies indicate that this is not the case. Fluorescent analogues of tubulin and actin have been invaluable reagents in understanding the dynamics of microtubules and microfilaments. Similar methods can be used to study IF dynamics. However, in the case of NFs, the modification of lysine residues by acylation abolishes their capacity to assemble (61), and very likely the same may be expected in other types of IFs. An alternative modification scheme exploits the fact, mentioned above, that IFsps usually contain few Cys residues, and that these do not normally form disulphide bonds under physiological conditions. In the case of bovine NF-L, there is a single Cys residue at position 321 and, in all the aspects analysed, the behaviour of sulphydryl-labelled NF-L is indistinguisable from the unmodified protein (61, see also 114). The fluorescent derivatization of bovine NF-L in the dissociated state is performed using the sulphydryl-specific fluorescent maleimides, 3-(4-maleimidylphenyl)-7-diethylamino-4-methylcoumarin and fluorescein-5-maleimide. The reaction mixture at a ratio of 1:10 protein:fluorophore in either 8 M urea, 150 mM sodium phosphate, pH 7.4, or 0.1 M MES, 0.17 M NaCl, pH 8.1, is maintained for 60 min at 22 °C. The reaction is terminated by addition of DTT (1 mM final concentration).

Dialysis for 12 h at 37 °C against 20 mM Hepes–Tris, pH 7.2, 10 mM NaCl, 1 mM MgCl₂ removes excess fluorophore and urea. Pelleting of IFs is avoided, since it reportedly produces extensive aggregation. Labelled NF-L is stored at 4 °C for no longer than 3 days. The same procedure is used to produce biotinylated NF-L using biotin–maleimide. The reaction is rapid and specific, and the final labelling stoichiometry approaches 1:1 under denaturing conditions. Interactions between IFs and IFsps can be measured using fluorescence quenching (61, 114).

4. Cell biology of IFs

4.1 Immunological techniques

In general, IFsps are quite immunogenic, and most standard immunization protocols (115) will generate a good immunogenic response. Because of their value for the diagnosis of tumour cell origins, a large number of anti-IFsp reagents, primarily murine monoclonal antibodies, have been made and are commercially available from many companies, including Sigma, Boehringer-Mannheim, and Amersham. However, for day-to-day research uses, and particularly for antibody-intensive projects such as antibody injection, the cost of purchasing these reagents can be prohibitive. Luckily, the number of hybridomas secreting well-defined anti-IFsp antibodies that are available from non-profit sources, in particular the NIH-supported Developmental Biology Hybridoma Bank (*Table 2*), continues to increase with time.

4.1.1 Using antibodies in immunofluorescence

There are only a few cautions that need to be mentioned in using anti-IF antibodies in immunofluorescence microscopy; the most important are the

Table 2. Available anti-IFsp-hybridoma lines

Clone code	Target IFsp	Species	Source
anti-IFA (TIB131)	IFsps	many species	ATCC[a]
TROMA-1	cytokeratin EndoA	mouse	DHB[b]
E/C8	neurofilament-associated protein	avian	DHB
D3	desmin	chick/rat	DHB
D76	desmin	chick	DHB
2H3	NFm	rat, mouse, chick	DHB
3A10	neurofilament-associated antigen	chick, rat, mouse	DHB
1h5	Type II cytokeratin	*Xenopus*	DHB
14a9	nuclear lamin	*Xenopus*	DHB
14h7	vimentin	*Xenopus*	DHB
RT97	neurofilament proteins	human, chick, rat	DHB

[a] American Type Cell Culture, 12301 Parklawn Drive, Rockville, MD 20852, USA.
[b] Developmental Studies Hybridoma Bank, c/o Dr T. August, Department of Pharmacology and Molecular Sciences, Johns Hopkins University School of Medicine, Biophysics Building, Room 311, 725 North Wolfe Street, Baltimore, MD 21205, USA.

problems of fixation-dependence and hidden epitopes. In general, many of the antibodies that are commercially available have been characterized with respect to their reactivity with formalin-fixed tissue-sections. This is particularly critical for work on medical pathology specimens. The methods for using anti-IFsp antibodies in immunocytochemistry of sections or cultured cells are relatively straightforward and similar to those used for other antibodies (see 115 and Chapter 1). The first step is to determine that the fixation to be used does not abolish immunoreactivity. This can best be done by testing the fixation on a cell that is known to stain with the antibody. We routinely use methanol and methanol–dimethylsuphoxide (4:1) solutions as fixatives. The methanol–dimethylsuphoxide fixative (Dent fixative) (116) works particularly well for whole-mount immunocytochemistry of *Xenopus* and other vertebrate embryos (117). Methanol-based fixatives are particularly good for IFsps which are already insoluble and do not require chemical cross-linking to maintain their localization within the cell or tissue.

If the antibody does not stain the specimen specifically, its epitope(s) may be hidden in the assembled IF. The existence of such hidden epitopes is not uncommon. Franke *et al.* (118) described a cell cycle- and fixation-dependent masking of an epitope on a type II cytokeratin. The epitope of this antibody could be unmasked in interphase cells by brief treatments with 2 M urea or trypsin after fixation, or by extraction of the cells with buffer containing 1 per cent Triton X-100 or Nonidet P-40 prior to fixation. We have found a similar masking of the monoclonal anti-cytokeratin antibody 1h5's epitope cultured *Xenopus* and mammalian cells, which can be unmasked by fixation for 5 min in 20 per cent acetic acid: 80 per cent methanol ($-20\ °C$). Similarly, the widely used monoclonal anti-cytokeratin antibodies (AE1 and AE3 (119), fail to stain *Xenopus* cytokeratins following methanol fixation, but stain well if specimens are treated briefly with 10 per cent hydrogen peroxide, which presumably 'opens up' IF structure in response to protein oxidation.

The molecular nature of differential epitope accessibility remains poorly understood. In some cases, IF-associated proteins could bind to the region and block antibody binding (e.g. 120). Alternatively, there may be changes in the accessibility of the epitope that result directly from the assembly of the subunit into an IF. For example, Birkenberger and Ip (121) used a monoclonal antibody that binds to the conserved 10–15 amino acid hydrophilic region of the tail domain of vimentin, desmin, GFAP, and cytokeratins 8 and 18 (122, 123). This monoclonal antibody binds to tetrameric desmin in solution and to monomeric desmin on Western blots, but fails to bind to desmin IFs. These data indicate that during assembly of an IF, this epitope becomes hidden.

4.1.2 Using antibodies in immunoblotting

Since its construction and characterization by Pruss *et al.* (124), the anti-IFA monoclonal antibody has become a widely used reagent in defining proteins as putative IFsps. Anti-IFA recognizes the IFA (intermediate filament antigen) epitope (TYRRLLEGE) located at the extreme C-terminal end of the central rod domain (125). However, care must be taken in the use of anti-IFA. A number of IFsps do not react with the antibody (see 87). Moreover, anti-IFA's ability to stain IFs in immunocytochemistry varies quite dramatically with the fixative used (124). In addition, there is also a real possibility that anti-IFA, like other monoclonal antibodies, may react with polypeptides unrelated to IFsps. The possibility of cross-reaction between IFsps and other proteins, particularly when using monoclonal antibodies, is due to the 'myopic' nature of monoclonal antibodies (126, 127). This is a problem with all studies using monoclonal antibodies, and is not unduly serious in the study of IFs.

The major technical problem associated with the use of anti-IFsp antibodies in immunoblotting is that many anti-cytokeratin antibodies, and a number of other antibodies, such as anti-IFA (124), react with human skin keratins. Since sloughed-off skin is a major component of household (and laboratory) dust, it is not uncommon to find human keratins as a contaminant of gel and transfer buffer solutions. Their complete elimination can be quite difficult, at least in our experience. This problem can be diagnosed by simply running blanks, consisting of samples prepared exactly as the real sample, but omitting the cells or tissue. The characteristic human skin keratins will be present. Typically, skin keratin contamination is not a significant problem when the target antigen is present in reasonable levels, but if trace amounts of the target are to be detected, the problem can become significant.

4.2 Studying IF assembly and organization dynamics *in vivo*

Two related methods have been used to study the assembly dynamics of IFsps *in vivo*, the intracellular injection of modified IFsps and the ectopic expression of IFsps. The first of these methods is best used with homopolymeric IFsps, because they are more soluble at low ionic strength than are the cytokeratins. In these studies, the purified IFsp is covalently modified (see above). Mittal *et al.* (128) labelled desmin with the fluorochrome iodoacetamidotetramethylrhodamine. Vikstrom *et al.* (129) labelled vimentin with biotin and then used a fluorescein-conjugated avidin to visualize the biotinylated vimentin. Similar experiments have been reported with the cytokeratins (130).

In a typical experiment, the modified IF protein is directly injected into cells, using a pressurized microneedle (see Chapter 1 and ref. 131). The distribution of the protein within the cell over time can be monitored either directly or indirectly by epifluorescence microscopy. Typically, the modified IFsp precipitates as it enters the cytoplasm, but over time the protein incorporates into the endogenous IF system. An alternative approach that avoids the potential artefacts associated with the initial precipitation of injected IFsps is the use of fluorescence recovery after photobleach (FRAP). In these studies the exogenous labelled protein is allowed to equilibrate within the cell for a number of hours after injection. Using a laser beam, small regions of the cell can then be bleached and the rate at which fluorescence signal returns to bleached area can be used to estimate the diffusion coefficient of the exogenous protein and the stability of IFsps within the IF (see Chapter 3 and 130). In these experiments, fluorochrome-conjugated IFsps can be used in much the same way that fluorescence-conjugated forms of α-actinin, actin, tubulin, and other cytoskeletal proteins have been used to study cytoskeletal dynamics. The same *caveats* that apply to interpretation of these studies apply to the use of FRAP in the study of IFs (see 132 for discussion).

An alternative approach to the study of IF assembly dynamics is to follow the appearance of newly synthesized IFsps in the cell. This requires either that the host cell not contain an IF system or that the newly synthesized IFsp be distinguishable from the IFsps of the host cell. While all eukaryotic cells appear to contain nuclear lamins, a number of cell lines that do not express cytoplasmic IF proteins have been described (5). An exogenous cytoplasmic IFsp expressed in IF⁻ cells can be recognized relatively simply (see ref. 113 for example). To visualize the interaction between an exogenous IFsp and a pre-existing IF system, it is necessary to distinguish exogenous and endogenous IFsps. Two methods can be used. The first involves using a different IFsp or an IFsp from a species different from the host cell. Antibodies specific for the exogenous IFsp must also be available. The second approach is to modify the IFsp by the addition of a 'tagging' sequence (see ref. 133). Albers and Fuchs (101, 102) used a sequence encoding the neuropeptide substance P; however, we have found this sequence difficult to use and the antibodies expensive. An alternative is the EQKLISEEDL(stop) sequence, derived from the human *c-myc* protein (134). This sequence is recognized specifically by the monoclonal antibody 9E10 (135); the 9E10 antibody is specific for this sequence and does not appear to react significantly with other proteins, including the *c-myc* proteins of other species. The sequence can be easily added to the 3' end of the coding sequence of any cloned cDNA. In this position, the tagging sequence does not appear to have any significant effect on the assembly of the neurofilament protein NF-L and NF-M (136), vimentin (107, Cary and Klymkowsky, unpublished observations), or cytokeratins (Bachant, Domingo and Klymkowsky, unpublished observations). The binding of 9E10 to its target sequence is inhibited by non-ionic detergents such as NP40, and detergents should be omitted from dilution and immunoblot wash buffers (Bachant, unpublished observations).

The immunologically distinguishable exogenous IFsp can be expressed in a cell in any of a number of ways. DNA encoding the protein can be transiently transfected using established procedures (137, 138). Alternatively, the sequence encoding the 'tagged' protein can be cloned behind an inducible promoter and stable lines can be prepared (see below). Upon addition of the inducer, the protein will be synthesized. An alternative method is to directly inject either *in vitro* synthesized mRNA into the cytoplasm or DNA into the nucleus of the cell. In the case of direct injection, the synthesis of the exogenous protein can be detected using immunofluorescence microscopy within 30 min to 1 h of injection (Klymkowsky *et al.*, unpublished observations). The procedure for direct intracellular or intranuclear injection is quite simple and makes use of the same equipment as antibody injection (see below).

4.3 Anti-IF reagents and the study of IF function

Despite an extensive search, drugs that specifically affect IF organization (and therefore presumably IF function) have not been discovered (4). However, a number of more complex reagents that appear *specifically* to disrupt IF organization or inhibit IFsp synthesis have been characterized. These include anti-IFsp and anti-IF-associated protein antibodies, anti-sense-mediated down-regulation, artificially constructed antimorphic mutant forms of IFsps and gene ablation. We will describe these methods, their strengths and their weaknesses in the context of studying IF organization and function.

4.3.1 Anti-IF antibodies

A wide range of anti-IFsp antibodies induce the reorganization of IFs when injected into a cell. Anti-IF-antibody-induced reorganization can vary from the collapse of the IF system toward the cell center to the fragmentation of IFs (see ref. 4 for a review). In our experience, if an anti-IF antibody binds to IFs in the cell, it will affect IF organization, provided its concentration is high enough. Typically, anti-IF effects are observed at concentrations above 1 mg/ml (injected solution) which, in microneedle-injected cells corresponds to an intracellular concentration of between 0.05 and 0.1 mg/ml. Microneedle injection, although apparently somewhat daunting, is really quite simple provided that the appropriate equipment is available (see Chapter 1). Alternative 'bulk' injection methods can be used, such as electroporation, osmotic shock, etc., but care must be taken with their use, since the delivery of antibody is not uniform.

Typically, antibody effects are obvious within an hour or two after injection, although some antibodies have been found to act within minutes of injection (Klymkowsky and Dent, unpublished observations; see 4). Monovalent Fab fragments act in much the same manner as intact antibody (139, Dent and Klymkowsky, unpublished observations), suggesting that the binding of the antibody to the IFsp rather than cross-linking effects between IFsps induces the change in IF organization. The time course of antibody effects is more difficult to estimate. We have found that antibody effects on IF organization can last from 12 to over 96 h, depending upon the antibody and the cell type. The transience of antibody effects can limit their usefulness for some studies. For example, Emerson (140) used antibody injection to disrupt cytokeratin filament organization in the early mouse embryo; the transience of the disruption, however, limited her studies to the earliest stages of development.

Where antibody injection produces a 'positive' effect, aside from the disruption of IF organization, there is concern, common to almost all experimental manipulations, that the observed effect is not specific. Assuming an initial antibody concentration of 5 mg/ml, intracellular antibody concentration can approach 500 µg/ml. At this concentration the antibody may react in a significant manner with epitopes that it would not affect at lower antibody concentration, such as are used for immunoblot (typically 1 to 5 µg/ml), or immunoprecipitation studies, where only relatively robust interactions will survive the washing steps. To control for this possibility, it is valuable to have a number of antibodies directed against non-overlapping epitopes on the target molecule. If each produces the same 'positive' effect, the likelihood that they are all reacting with non-target proteins to produce that effect is greatly reduced.

4.3.2 Anti-sense reagents

The selection and characterization of mutations in specific loci has been a traditional and useful approach to study protein function. Unfortunately, standard genetics has yet to provide much insight into IF function. This has spurred the development of artificial mutations that affect IFs. The concept of using anti-sense RNA to depress the synthesis of specific proteins originated with the observation that RNA sequences complementary to specific mRNAs interfered with the translation of those mRNAs (141, 142; see 143–145 for reviews). If a cDNA sequence for the IFsp of interest has been cloned, anti-sense reagents can be constructed by simply reversing the orientation of the cDNA with respect to a promoter. Transcription then uses the non-coding strand of the DNA as a template and produces an RNA complementary (anti-sense) to the normal (sense) mRNA. Sense and anti-sense RNAs form double-stranded RNA hybrids that appear to interfere with mRNA transcription (146), processing (147), cytoplasmic transport (148), and translation (149), resulting in a specific inhibition of the synthesis of the protein product of the target gene.

Although the appeal of anti-sense reagents lies in their simplicity and theoretical specificity for specific targets, there is actually very little known about the possible non-specific effects of high levels of anti-sense and double-stranded RNA on gene expression. This is a concern of particular significance in model systems of development, where even subtle non-specific effects on gene expression may have profound effects. Inhibition in the expression of a variety of cellular proteins (150), including other cytoskeletal constituents

(151, 152), has been achieved using anti-sense RNA expression vectors. IFs do pose a challenge to anti-sense manipulation. IFsps are abundant (153) and long lived (154). Although a rigorous analysis has not been performed, IF mRNAs appear to be both stable (155) and abundant (156). Consequently, a high level of anti-sense-RNA expression must be produced over a period of days to affect significantly the IFsp content of cells. Nevertheless, a significant reduction in the cellular content of cytokeratin IFsp has been observed (157), and there are preliminary reports indicating reduced desmin (158) and GFAP (159) in cells transfected with IF antisense expression vectors.

i. IF anti-sense gene constructs
Based on the understanding of anti-sense RNA and the experience of two of the present authors (A. S. and R. E.) with vimentin anti-sense vectors, it is possible to define the desirable features of an IF-anti-sense vector. Secondary as well as primary structure of target mRNAs can influence the efficiency of anti-sense RNA inhibition (160, 161). To avoid potential secondary structure interference, it is advisable to use less than full-length anti-sense cDNAs coding for relatively A–T-enriched regions. Recent studies have also suggested that the stability of anti-sense transcripts can be increased by the construction of chimeric cDNAs containing the anti-sense sequences joined to the coding sequence for structural RNAs (162, 143, 144). For abundant transcription, like that of IFsps, it is important to use a strong promoter to drive anti-sense RNA transcription. Strong promoter sequences such as the MSV- (163), SV40- (146), β-actin- (152), RSV long terminal repeat- (164), and MMTV-promoters (165) have all been used successfully in anti-sense expression plasmids. Inducible promoters, such as the hormone response element of MMTV (112, 113), may be preferred over promoters that are constitutively active. The main consideration in selecting an inducible promoter is the degree to which the inducing agent may affect the phenotype to be studied.

To produce stable transfectants, cells are transfected with both the anti-sense vector of interest and a gene that allows for the positive selection of those cells that received and express the exogenous DNAs. Selectable marker genes of bacterial origin can be used in any eukaryotic cell without prior selection, although the antibiotics used in the selective medium are relatively expensive. Currently, the most widely used selectable marker gene is probably aminoglycoside phosphotransferase (APH or *neo*), which confers resistance to the antibiotic G-418. *Neo*-containing plasmids are widely available (see 138).

ii. Episomal vs. chromosomal integration plasmids
Expression plasmids that require integration into the host cell chromosome are the most commonly used to produce stable transfectants. There are now a number of well-characterized expression plasmids with multiple cloning sites

244

and promoters with wide host ranges. Marker genes can be directly included in the expression plasmid itself or introduced into the host cell, together with the anti-sense expression plasmid, by co-transfection. In co-transfection, the anti-sense expression plasmid is usually added at 10–40 times the marker gene-containing plasmid. Co-transfection results in fewer stable clones per µg of transfected DNA and fewer clones per transfection, compared with transfections using expression plasmids that contain the selectable gene.

Expression plasmids that replicate episomally in transfected cells contain the *cis*-acting replication element *ori-P* of Epstein–Barr virus (EBv) (166) together with a selectable marker. Such plasmids have been constructed that contain the metallothionein and Rous sarcoma virus 3′ long terminal repeat promoters (pREP vectors) (164). Since they do not normally integrate into host cell chromosomes, episomal plasmids generally do not cause the non-specific mutagenesis that can be associated with plasmid insertion into genomic DNA. In addition, plasmid copy number is sensitive to the concentration of the selectable agent in the culture medium. Higher levels of expression can be selected by increasing the concentration of the selecting agent. On the other hand, the host species range of EBv *ori-P*-based vectors may be limited; particularly in non-human cells there is some question of the degree to which these plasmids remain episomal. There also appear to be limits to the copy number of plasmids per cell that can be obtained, and they may be less effective than integration plasmids in producing high levels of anti-sense RNA.

iii. Transfection, selection, and screening of stable cell lines with an IF anti-sense phenotype

The abundance and stability of IFsps and their mRNAs make transient transfections of limited value in IF anti-sense expression experiments. Instead, stable cell lines that express IF anti-sense RNA must be produced. This leads to a steady state between sense and anti-sense IF RNA expression. If a sufficiently high level of anti-sense RNA is expressed, the effects of reduced IFsp synthesis can be studied over time. In an anti-sense experiment, the screening of the largest number of individual transfected clones that can be practically obtained is important for at least two reasons. First, experiments in which MMTV-inducible IF cDNAs were introduced into cultured cell lines find that only a fraction (10–15 per cent) of the stable G-418 resistant clones obtained expressed detectable levels of IFsp (112; Sarria *et al.*, unpublished observations). Most likely this is because the level of aminoglycoside phosphotransferase expression required for G-418 resistance is significantly lower than the level of IFsp expression required for IFsp detection in immunofluorescence microscopy. Second, for reasons that are less clear, in many transfectant clones we find that the expression of IFsps is quite heterogenous. The fraction of the cells that contain detectable IFsps can vary considerably between individual clones or even subclones. These observations indicate that stable transfectants that uniformly express high levels of the transfected gene do not occur with high frequency. Therefore, expression plasmids, host cells, and transfection protocols that yield large numbers of stable clones are important considerations.

Plasmid DNA to be transfected into cells needs to be as highly purified as possible. Plasmid DNA preparations should be banded at least once, and preferably twice, on caesium chloride gradients. Purified plasmid DNA is concentrated and sterilized prior to transfection by ethanol precipitation. The most common techniques for stable transfection of mammalian cells include calcium phosphate precipitation (167), lipofection (168), and electroporation (169). Calcium phosphate transfection is probably the most widely used; it is less expensive than either lipofection or electroporation, but the number of colonies obtained is often not as reproducible. Lipofection requires the cells to be in a serum-free medium which may not be tolerated well by some cell types. For all transfection protocols, transfection efficiency is dependent on DNA concentration, and different cells can differ significantly with respect to viability and transfectability. It is advisable to conduct preliminary experiments to determine an optimal DNA concentration and, in the case of calcium phosphate or lipofection protocols, the length of time that the cells will tolerate the transfection conditions. Following transfection, the cells should be maintained in a subconfluent condition for 2 to 3 days and then placed in medium containing the selective agent. Depending on the ability of a given cell type to grow from low density, stable transfectant colonies can be recovered within 2 to 3 weeks. Because of the large numbers of clones that can be generated from a single transfection, clones should be screened for an anti-sense phenotype as soon as possible. The use of indirect immuno-fluorescence to screen individual transfectants for a reduced IF content is a reasonable initial step before further characterization of selected clones. Clones that appear to have a reduced IF content must be further characterized by Northern blotting and SDS-PAGE for sense and anti-sense RNA and IFsp content.

4.3.3. Artificial antimorphic mutations

Because IFs are oligomeric structures, it is possible to construct mutant IFsps that will interact with native IFsps and 'poison' their ability to form IFs. Such mutants appear analogous to the 'antimorphic' mutations defined genetically by Muller (170). Herskowitz (171) refers to such mutants as dominant negative-effect mutations. Albers and Fuchs (101) constructed the first antimorphic mutant of an IFsp using a type I cytokeratin (see also 172, 100, 173). Since then, antimorphic mutations have been constructed for type II cytokeratins (100; Bachant, Domingo and Klymkowsky, in preparation), nuclear lamins (174), the neurofilament proteins NF-L and NF-M (105, 106), vimentin (107, Cary and Klymkowsky, unpublished observations) and desmin (104). In most cases these mutants involve the truncation of the protein to

Alberto Domingo et al.

remove the C-terminal tail domain together with the IFA epitope. As the truncations become longer, the antimorphic nature of the mutant protein becomes more severe. Similar truncations of the N-terminal have a more complex behaviour. In the type I cytokeratin analysed by Albers and Fuchs (102), they appear to be potent antimorphic mutations; in contrast in the neurofilament proteins (105, 106), desmin (104), and vimentin (Cary and Klymkowsky, unpublished observations) similarly N-terminally 'decapitated' proteins have no effect on IF organization. Mutant IFsps can be constructed through the use of convenient restriction sites, digestion of linearized plasmids with exonuclease, or the polymerase chain reaction (*Figure 1*). In addition to deletions from the ends, Letai et al. (175) have found that point mutations in the IFA epitope region of a type I cytokeratin results in an antimorphic mutant polypeptide.

Antimorphic mutant IF proteins can be introduced into cells in a number of ways. First, plasmids encoding the mutant proteins can be introduced using transient transfection with the methods described in Ausubel *et al.* (137) or Sambrook *et al.* (138); this enables the effect of the mutant on IF organization to be examined. To examine the long-term effects of mutant proteins, cells can be stably transformed with plasmids encoding the mutant proteins, under constitutive or inducible promoter control, using the same methods as described for introducing anti-sense plasmids (see above). In *Xenopus* oocytes or embryos, mutant proteins can be synthesized by injecting either *in vitro*-synthesized mRNA or plasmid DNA encoding the mutant. RNA has a half-life of approximately 10 h in the embryo and so can be used to study events up through neurulation (20 to 30 h after fertilization) (see 107). In contrast, DNA is expressed following the midblastula transition and into the swimming tadpole stages of development. DNA is expressed in a highly mosiac manner, however, and so in general the effect of the mutant protein on the structure and behaviour of individual cells, rather than whole tissues, can be examined. In the mouse embryo, DNA is introduced into the male pronucleus to construct transgenic mice (see 176).

The recent excitement about antimorphic mutants in IFsps is that under appropriate promoter control, these proteins can be synthesized in specific tissues. This approach has been used with antimorphic cytokeratin mutations in a differentiating embryonal carcinoma cell line (173) and most impressively in transgenic mice (177). Transgenic homozygous mice expressing an antimorphic mutant type I cytokeratin in the basal layer of the epidermis exhibit a severe mutant phenotype resembling the human skin dieases epidermolysis bullosa simplex (177). This dramatic result leads inevitably to the question of what kind of artefacts the expression of antimorphic IFsp mutants can produce. This is a real concern, since in a typical ectopic expression experiment the level of the mutant polypeptide is usually high and is, after all, a mutant polypeptide with an altered structure. Its interactions with other cellular structures may be abnormal as well. Also, since the

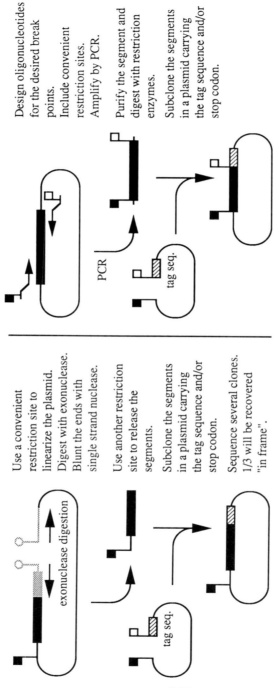

Design oligonucleotides for the desired break points.
Include convenient restriction sites.
Amplify by PCR.

Purify the segment and digest with restriction enzymes.

Subclone the segments in a plasmid carrying the tag sequence and/or stop codon.

PCR

tag seq.

Use a convenient restriction site to linearize the plasmid.
Digest with exonuclease.
Blunt the ends with single strand nuclease.

Use another restriction site to release the segments.

Subclone the segments in a plasmid carrying the tag sequence and/or stop codon.

Sequence several clones. 1/3 will be recovered "in frame".

exonuclease digestion

tag seq.

248

Figure 1. Two general ways to make mutant IFsps. In the first (left) the plasmid is linearized 'down-stream' of the coding sequence and digested with exonuclease. The extent of digestion determines the length of the deletion. Subcloning and sequencing fragments permit the identification of 'in-frame' mutants. In the right-hand panel, oligonucleotides are used to select the sites of mutation. Polymerase chain reaction (PCR) is then used to amplify the selected region, and the amplified fragment is then subcloned into an appropriate vector. Care must be taken, since PCR amplification can introduce mutations into the amplified sequence.

synthesis of the mutant protein is typically restricted to a specific tissue, the phenotype observed will, whether specific or not, almost certainly be observed in that tissue. Studies of antimorphic mutants of cytokeratin indicate that certain of these mutants can interact aberrantly with other cellular proteins. For example, the NΔ117pJK14-P mutant of the type I human cytokeratin K14 is found within the nucleus of cells (102), a localization that strongly suggests that it interacts with karyophilic components. As in the case of injected antibodies, there is a real possibility that the mutant protein may interact in an unanticipated manner with other cellular components, and that it is this interaction, rather that the interaction with the endogenous IF system, that produces the mutant phenotype. To control for this possibility is difficult, but perhaps the best approach is to determine whether there is a strict correlation between the ability of a series of mutants to disrupt IF organization and the observed cellular or tissue phenotype they produce.

4.4 Ectopic expression of IFsps

The tight, tissue-specific regulation of IFsps suggested that their ectopic expression might cause changes in cellular behaviour. This type of experiment has now been performed in both cultured cells and in transgenic mice (reviewed in ref. 4). To date, no specific changes in cellular or tissue behaviour have been noted, save for defects in the lens of the eye (178, 179) and cyclic hair-loss (180). Based on these experiments, it appears that the ectopic, and often unregulated, expression of IFsps can disrupt normal cellular functions, but that it provides little new information as to the functions of IFs.

4.5 Gene ablation

The final approach to the study of IF function is the construction of embryonic stem cell lines in which specific endogenous IF genes have been disrupted by homologous recombination (see 181 for review). Such mutant cell lines have been made for the cytokeratin ENDO A (182). The mutant cells can then be incorporated into mouse embryos and occasional germ line transformants can, at least in theory, be established. Such mice can then be bred to produce homozygous mutant mice that completely lack the initially targeted IFsp. The phenotypes of these animals should reflect defects arising from the absence of the specific IFsp. Although there is a possibility the other IFsps could partially compensate for the missing IFsp, the tight cell-type expression of IF proteins, and the lack of transcriptional cross-regulation suggest that in most cases this will not be a problem.

5. Conclusion

The study of IF organization and function combines relatively simple biochemical techniques with sophisticated immunological and molecular biological approaches to address an intriguing question: what functions do IFs perform? As reagents, i.e., monoclonal antibodies, tagged IFsps, and antimorphic mutant IFsps, proliferate, the challenge will be to define the specificity of their effects and to use them in appropriate context to study the role of IFs in normal tissue function.

Acknowledgements

We thank the NIH (RME), the NSF and Pew Scholars Program (MWK) for their support of our research. AD is a postdoctoral fellow of the Ministerio de Educatión y Cinecia of Spain.

References

1. Traub, P. (1985). *Intermediate filaments*. Springer-Verlag, Berlin.
2. Duboulay, C. E. H. (1985). *J. Pathol.*, **146**, 77.
3. Steinert, P. M. and Roop, R. (1988). *Ann. Rev. Biochem.*, **57**, 593.
4. Klymkowsky, M. W., Bachant, J. B., and Domingo, A. (1989). *Cell Motil. Cytoskeleton*, **14**, 309.
5. Sarria, A. J. and Evans, R. M. (1989). *RBC Rev.*, **22**, 1.
6. Trask, D. K., Band, V., Zajchowski, D. A., Yaswn, P., Suh, T., and Sager, R. (1990). *Proc. Natl. Acad. Sci. USA*, **87**, 2319.
7. Bartnik, E. and Weber, K. (1989). *Eur. J. Cell Biol.*, **50**, 17.
8. Chang, X. and Piperno, G. (1987). *J. Cell Biol.*, **104**, 1563.
9. Ishikawa, H., Bishoff, R., and Holtzer, H. (1968). *J. Cell Biol.*, **38**, 538.
10. McKeon, F. D., Kirschner, M. W., and Caput, D. (1986). *Nature*, **319**, 463.
11. Fisher, D. Z., Chaudhary, N., and Blobel, G. (1986). *Proc. Natl. Acad. Sci. USA*, **83**, 6450.
12. Pachter, J. S. and Liem, R. K. H. (1985). *J. Cell Biol.*, **101**, 1316.
13. Fliegner, K. H., Ching, C. Y., and Liem, R. K. H. (1990). *EMBO J.*, **9**, 749.
14. Leonard, D. G. B., Gorham, J. D., Cole, P., Greene, L. A., and Ziff, E. B. (1988). *J. Cell Biol.*, **106**, 181.
15. Landon, F., Lemonnier, M., Benarous, R., Huc, C., Fiszman, M., Gros, F., and Portier, M. M. (1989). *EMBO J.*, **8**, 1719.
16. Lendahl, U., Zimmerman, L. B., and McKay, R. D. G. (1990). *Cell*, **60**, 585.
17. Shaw, G. and Hawkins, J. (1990). *J. Cell Biol.*, **111**, 42 (abstract).
18. Remington, S. G. (1990). *J. Cell Biol.*, **111**, 44 (abstract).
19. Capetanaki, Y., Kusik, I., and Rothblum, K. (1990). *Oncogene*, **5**, 645.
20. Cohen, C. and Parry, D. A. D. (1986). *TIBS*, **11**, 245.
21. Weber, K., Plessmann, U., Dodemont, H., and Kossmagk-Stephan, K. (1988). *EMBO J.*, **7**, 2995.
22. Weber, K., Plessmann, U., and Ulrich, W. (1989). *EMBO J.*, **8**, 3221.
23. Moll, R., Franke, W. W., Schiller, D. L., Geiger, B., and Kreplet, R. (1982). *Cell*, **31**, 11.
24. Sun, T. T. and Green, H. (1978). *J. Biol. Chem.*, **253**, 2053.

25. Weber, K. and Giesler, N. *Intermediate filaments from wool α-keratin to neurofilaments: a structural overview*, p. 153. Cold Spring Harbor Laboratory, New York.
26. Steinert, P. M., Idler, W. W., and Goldman, R. D. (1980). *Proc. Natl. Acad. Sci. USA*, **77**, 4534.
27. Geisler, N. and Weber, K. (1983). *EMBO J.*, **2**, 2059.
28. Mori, H., Kondo, J., and Ihara, Y. (1987). *Science*, **235**, 164.
29. Ohta, M., Marceau, N., Perry, G., Maneto, V., Gambetti, P., Autilio-Gambetti, L., Metuzals, J., Cadrin, H. K. M., and French, S. W. (1988). *Lab. Invest.*, **59**, 848.
30. Gard, D. L. and Lazarides, E. (1979). *J. Cell Biol.*, **81**, 336.
31. Beck, L. A., Hosick, T. J., and Sinensky, M. (1988). *J. Cell Biol.*, **107**, 1307.
32. Beck, L. A., Hosick, T. J., and Sinensky, M. (1990). *J. Cell Biol.*, **110**, 1489.
33. Wolda, S. L. and Glomset, J. A. (1988). *J. Biol. Chem.*, **263**, 5997.
34. Vidrich, A., Gilmartin, M., Zimmerman, J., and Freedberg, I. M. (1982). *J. Cell Biol.*, **95**, 237a.
35. Baden, H. P. and Lee, L. D. (1978). *J. Invest. Dermatol.*, **71**, 141.
36. Steinert, P. M. and Idler, W. W. (1979). *Biochemistry*, **18**, 5664.
37. Wong, J., Hutchison, S. B., and Liem, R. K. H. (1984). *J. Biol. Chem.*, **259**, 10867.
38. Carden, M. J., Schlaepfer, W. W., and Lee, V. M. Y. (1985). *J. Biol. Chem.*, **260**, 9805.
39. Heald, R. and McKeon, F. (1990). *Cell*, **61**, 579.
40. Evans, R. M. and Fink, L. M. (1982). *Cell*, **29**, 43.
41. Celis, J. E., Mose-Larsen, P., Fey, S. J., and Celis, A. (1983). *J. Cell Biol.*, **97**, 1429.
42. Evans, R. M. (1988). *Eur. J. Cell Biol.*, **46**, 152.
43. Peter, M., Nakagawa, J., Doree, M., Labbe, J. C., and Nigg, E. A. (1990). *Cell*, **61**, 591.
44. Ward, G. E. and Kirschner, M. W. (1990). *Cell*, **61**, 561.
45. Chou, Y. H., Bischoff, J. R., Beach, D., and Goldman, R. D. (1990). *Cell*, **62**, 1063.
46. Klymkowsky, M. W. and Maynell, L. A. (1989). *Devel. Biol.*, **134**, 479.
47. Nixon, R. A. and Lewis, S. E. (1986). *J. Biol. Chem.*, **261**, 16298.
48. Nixon, R. A., Lewis, S. E., and Marotta, C. A. (1987). *J. Neurosci.*, **7**, 1145.
49. O'Conner, C. M., Gard, D. L., and Lazarides, E. (1981). *Cell*, **23**, 135.
50. Toru-Delbauffe, D., Pierre, M., Osty, J., Chantoux, F., and Francon, J. (1986). *Biochem. J.*, **235**, 283.
51. Inagaki, M., Nishi, Y., Nishizawa, K., Matsuyama, M., and Sato, C. (1987). *Nature*, **328**, 649.
52. Chou, Y. H., Rosevear, E., and Goldman, R. D. (1989). *Proc. Natl. Acad. Sci. USA*, **86**, 1885.
53. Evans, R. M. (1989). *J. Cell Biol.*, **108**, 67.
54. Wible, B. A., Smith, K. E., and Angelides, K. J. (1989). *Proc. Natl. Acad. Sci. USA*, **86**, 720.
55. Evans, R. M. (1988). *FEBS Lett.*, **234**, 73.
56. Inagaki, M., Gonda, Y., Matsuyama, M., Nishizawa, K., Nishi, Y., and Sato, C. (1988). *J. Biol. Chem.*, **263**, 5970.

57. Geisler, N. and Weber, K. (1988). *EMBO J.*, **7**, 15.
58. Ando, S., Tanake, K., Gonda, Y., Sato, C., and Inagaki, M. (1989). *Biochemistry*, **28**, 2974.
59. Yeagle, P. L., Frye, J., and Eckert, B. S. (1990). *Biochemistry*, **29**, 1508.
60. Eckert, B. S. and Yeagle, P. L. (1988). *Cell Motil. Cytoskeleton*, **11**, 24.
61. Angelides, K. J., Smith, K. E., and Takeda, M. (1989). *J. Cell Biol.*, **108**, 1495.
62. Lamb, N. J. C., Fernandez, A., Feramisco, J. R., and Welch, W. J. (1989). *J. Cell Biol.*, **108**, 2409.
63. Bravo, R., Small, J. V., Fey, S., Larsen, P. M., and Celis, J. E. (1982). *J. Mol. Biol.*, **154**, 121.
64. Soellner, P., Quinlan, R. A., and Franke, W. W. (1985). *Proc. Natl. Acad. Sci. USA*, **80** 993.
65. Isaacs, W. B., Cook, R. K., Atta, J. C. V., Redmond, C. M., and Fulton, A. B. (1990). *J. Biol. Chem.*, **264**, 17953.
66. Nelson, W. J. and Traub, P. (1982). *J. Cell Sci.*, **57**, 25.
67. Nelson, W. J. and Traub, P. (1983). *Mol. Cell Biol.*, **3**, 1146.
68. Vorgias, C. E. and Traub, P. (1986). *Biosci. Rep.*, **6**, 57.
69. Fischer, S., Vandekerckhove, J., Ampe, C., Traub, P., and Weber, K. (1986). *Biol. Chem. Hoppe-Seyler*, **367**, 1147.
70. Schlaepfer, W. W. and Zimmerman, U. J. P. (1985). *Ann. N.Y. Acad. Sci.*, **455**, 552.
71. Schlaepfer, W. W. (1987). *J. Neuropath. Exp. Neurol.*, **45**, 117.
72. Lorand, L., Conrad, S. M., and Velasco, P. T. (1985). *Ann. N.Y. Acad. Sci.*, **455**, 703.
73. Stewart, M. (1990). *Curr. Opinions Cell Biol.*, **2**, 91.
74. Brice, C. F. C. and Crichton, R. R. (1971). *J. Chromatography*, **63**, 267.
75. Fish, W. W., Mann, K. G., and Tanford, C. (1989). *J. Biol. Chem.*, **144**, 4989.
76. Franke, W. W., Winter, S., Schmid, E., Soellner, P., Hammerling, G., and Achstatter, T. (1987). *Exp. Cell Res.*, **173**, 17.
77. Quinlan, R. A., Cohlberg, J. A., Schiller, D. L., Hatzfeld, M., and Franke, W. W. (1984). *J. Mol. Biol.*, **178**, 365.
78. Traub, P., Perides, G., Scherbarth, A., and Traub, U. (1985). *FEBS Lett.*, **193**, 217.
79. Traub, P., Perides, G., Schimmel, H., and Scherbarth, A. (1986). *J. Biol. Chem.*, **261**, 10558.
80. Traub, P., Perides, G., Kuhn, S., and Scherbarth, A. (1987). *Eur. J. Cell Biol.*, **43**, 55.
81. Traub, P. and Nelson, W. J. (1983). *Hoppe Seyler's Z. Physiol. Chem.*, **364**, 575.
82. Traub, P. and Vorgias, C. E. (1984). *J. Cell Sci.*, **65**, 1.
83. Traub, P., Nelson, W. J., Kuhn, S., and Vorgias, C. E. (1982). *J. Biol. Chem.*, **258**, 1456.
84. Traub, P., Nelson, W. J., Kuhn, S., and Vorgias, C. E. (1983). *J. Biol. Chem.*, **258**, 1456.
85. Traub, P., Vorgias, C. E., and Nelson, W. J. (1985). *Mol. Biol. Rep.*, **10**, 129.
86. Shoeman, R. L., Wadle, S., Scherbarth, A., and Traub, P. (1988). *J. Biol. Chem.*, **263**, 18744.
87. Magin, T. M., Hatzfeld, M., and Franke, W. W. (1987). *EMBO J.*, **6**, 2607.
88. Stewart, M., Quilan, R. A., and Moir, R. D. (1989). *J. Cell Biol.*, **109**, 225.

89. Coulombe, P. and Fuchs, E. (1990). *J. Cell Biol.*, **111**, 153.
90. Nagai, K. and Thorgersen, H. C. (1987). *Meth. Enzymol.*, **153**, 461.
91. Nelson, W. J. and Traub, P. (1982). *J. Biol. Chem.*, **257**, 5536.
92. Nelson, W. J., Vorgias, C. E., and Traub, P. (1982). *Biochem. Biophys. Res. Comm.*, **106**, 1141.
93. Shoeman, R. L. and Traub, P. (1990). *J. Biol. Chem.*, **265**, 9055.
94. Evans, R. M. (1984). *J. Biol. Chem.*, **259**, 5372.
95. Hunkapiller, M. W., Lujan, E., Ostrander, F., and Hood, L. E. (1983). *Meth. Enzymol.*, **91**, 227.
96. Hatzfeld, M. and K. Weber, (1990). *J. Cell Biol.*, **110**, 1199.
97. Earnshaw, W. C., Honda, B. M., Laskey, R. A., and Thomas, J. O. (1980). *Cell*, **21**, 373.
98. Bader, B. L., Magin, T. M., Hatzfeld, M., and Franke, W. W. (1986). *EMBO J.*, **5**, 1865.
99. Hatzfeld, M. and Weber, K. (1990). *J. Cell. Sci.*, **97**, 317.
100. Lu, X. and Lane, E. B. (1990). *Cell*, **62**, 681.
101. Albers, K. and Fuchs, E. (1987). *J. Cell Biol.*, **105**, 245.
102. Albers, K. and Fuchs, E. (1989). *J. Cell Biol.*, **108**, 1477.
103. Geisler, N., Kaufmann, E., and Weber, K. (1982). *Cell.*, **30**, 277.
104. Raats, J. M. H., Piper, F. R., Egberts, W. T. M. V., Verrijp, K. N., Ramaekers, F. C. S., and Bloemendal, H. (1990). *J. Cell Biol.*, **111**, 1971.
105. Gill, S. R., Wong, P. C., Monteiro, M. J., and Cleveland, D. W. (1990). *J. Cell Biol.*, **111**, 2005.
106. Wong, P. C. and Cleveland, D. W. (1990). *J. Cell Biol.*, **111**, 1987.
107. Christian, J. L., Edelstein, N. G., and Moon, R. T. (1990). *The New Biol.*, **2**, 700.
108. Zackroff, R. V. and Goldman, R. D. (1979). *Proc. Natl. Acad. Sci. USA*, **76**, 6226.
109. Renner, W., Franke, W. W., Schmid, E., Giesler, N., Weber, K., and Mandelkow, E. (1981). *J. Mol. Biol.*, **149**, 285.
110. Steinert, P. M., Idler, W. W., Cabral, F., Gottesman, M. M., and Goldman, R. D. (1981). *Proc. Natl. Acad. Sci. USA*, **78**, 3692.
111. Fraser, R. D. B., Macrae, T. P., Parry, D. A. D., and Suzuki, E. (1986). *Proc. Natl. Acad. Sci. USA*, **83**, 1179.
112. Ngai, J., Coleman, T. R., and Lazarides, E. (1990). *Cell*, **60**, 415.
113. Sarria, A. J., Nordeen, S. K., and Evans, R. M. (1990). *J. Cell Biol.*, **111**, 553.
114. Ip, W. and Fellows, M. E. (1989). *Anal. Biochem.*, **185**, 10.
115. Harlow, E. and Lane, D. (1988). *Antibodies*. Cold Spring Harbor Press, New York.
116. Dent, J. A., Polson, A. G., and Klymkowsky, M. W. (1989). *Development*, **105**, 61.
117. Klymkowsky, M. W. and Hanken, J. (1991). *Meth. Cell Biol.*, **36**, 413.
118. Franke, W. W., Schmid, E., Wellsteed, J., Grund, C., Gig, O., and Geiger, B. (1983). *J. Cell Biol.*, **97**, 1255.
119. Tseng, S. C. G., Jarvin, M. J., Nelson, W. G., Huang, J. H., Woodcock-Mitchell, J., and Sun, T. T. (1982). *Cell*, **30**, 361.
120. Eckert, B. S., Koons, S. J., Schantz, A. W., and Zoblel, C. R. (1980). *J. Cell. Biol.*, **86**, 1.

121. Birkenberger, L. and Ip, W. (1990). *J. Cell Biol.*, **111**, 2063.
122. Franke, W. W. (1987). *Cell Biol. Int. Rep.*, **11**, 831.
123. Krauss, S. and Franke, W. W. (1990). *Gene*, **86**, 241.
124. Pruss, R. M., Mirsky, R. M., Raff, M. C., Thorpe, R., Dowling, A. J., and Anderton, B. H. (1981). *Cell*, **27**, 419.
125. Geisler, N., Kaufmann, E., Fischer, S., Plessmann, U., and Weber, K. (1983). *EMBO J.*, **2**, 1295.
126. Richards, F. F., Konigsberg, W. H., Rosenstein, R. W., and Vaga, J. M. (1975). *Science*, **187**, 130.
127. Lane, D. P. and Koprowski, H. (1982). *Nature*, **296**, 200.
128. Mittal, B., Sanger, J. M., and Sanger, J. W. (1989). *Cell Motil. Cytoskeleton*, **12**, 127.
129. Vikstrom, K. L., Borisy, G. G., and Goldman, R. D. (1989). *Proc. Natl. Acad. Sci. USA*, **86**, 549.
130. Vikstrom, K. L., Lim, S. S., Borisy, G. G., and Goldman, R. D. (1990). *J. Cell Biol.*, **111**, 260 (abstract).
131. Kreis, T. E. and Birchmeier, W. (1982). *Int. Rev. Cytol.*, **75**, 209.
132. Vigers, G. P. A., Coue, M., and McIntosh, J. R. (1988). *J. Cell Biol.*, **107**, 1011.
133. Munro, S. and Pelham, H. R. B. (1984). *EMBO J.*, **3**, 3087.
134. Munro, S. and Pelham, H. R. B. (1987). *Cell*, **48**, 899.
135. Evan, G. I., Lewis, G. K., Ramsey, G., and Bishop, J. M. (1985). *Mol. Cell. Biol.*, **5**, 3610.
136. Monteiro, M. J. and Cleveland, D. W. (1989). *J. Cell Biol.*, **108**, 579.
137. Ausubel, F. M., Brent, R., Kingston, R. E., Moore, D. D., Seidman, J. G., Smith, J. A., and Struhl, K. (1987). *Current protocols in molecular biology*. Wiley, New York.
138. Sambrook, J., Fritsh, E. F., and Maniatis, T. (1989). *Molecular cloning: a laboratory manual*. Cold Spring Harbor Press, New York.
139. Tolle, H. G., Weber, K., and Osborn, M. (1986). *Exp. Cell Res.*, **162**, 462.
140. Emerson, J. A. (1988). *Development*, **104**, 219.
141. Paterson, B. M., Roberts, B. E., and Kuff, E. L. (1977). *Proc. Natl. Acad. Sci. USA*, **74**, 4370.
142. Hastie, N. D. and Held, W. A. (1978). *Proc. Natl. Acad. Sci. USA*, **75**, 1217.
143. Izant, J. G. (1989). *Cell. Motil. Cytoskeleton*, **14**, 81.
144. Helene, C. and Toulme, J. J. (1990). *Biochim. Biophys. Acta*, **1049**, 99.
145. Colman, A. (1990). *J. Cell Sci.*, **97**, 399.
146. Yokoyama, K. and Imamoto, F. (1987). *Proc. Natl. Acad. Sci. USA*, **84**, 7363.
147. Munroe, S. H. (1988). *EMBO J.*, **7**, 2523.
148. Kim, S. K. and Wold, B. J. (1985). *Cell*, **42**, 129.
149. Melton, D. A. (1985). *Proc. Natl. Acad. Sci. USA.*, **82**, 144.
150. van der Krol, A. R., Mol, J. N. M., and Stuitje, A. R. (1988). *Biotech.*, **6**, 958.
151. Izant, J. G. and Weintraub, H. (1985). *Science*, **229**, 345.
152. Gunning, P., Leavitt, J., Muscat, G., Ng, S. Y., and Kedes, L. (1987). *Proc. Natl. Acad. Sci. USA*, **84**, 4831.
153. Bravo, R. and Celis, J. E. (1982). *Clin. Chem.*, **28**, 766.
154. Cabral, F. and Gottesman, M. M. (1979). *J. Biol. Chem.*, **254**, 6203.
155. Lilienbaum, A., Legagneux, U., Portier, M. M., Dellagi, K., and Paulin, D. (1986). *EMBO J.*, **5**, 2809.

156. Kreis, T. E., Geiger, B., Schmid, E., Jorcano, J. L., and Franke, W. W. (1983). *Cell*, **32**, 1125.
157. Trevor, K., Linney, E., and Oshima, R. G. (1987). *Proc. Natl. Acad. Sci. USA*, **84**. 1040.
158. Choudhary, S. K. and Capetanaki, Y. (1990). *J. Cell Biol.*, **111**, 44 (abstract).
159. Chen, W. J., Weinstein, D. E., and Liem, R. K. H. (1990). *J. Cell Biol.*, **111**, 43 (abstract).
160. Cavanaugh, P. F. and Pilgermayer, W. F. (1989). *Cancer Res. (Proceedings)* **30**, 417 (abstract).
161. Verspieren, P., Loreau, N., Thuong, N. T., Shire, D., and Toulme, J. J. (1990). *Nucl. Acid. Res.*, **18**, 4711.
162. Takayama, K. M. and Inouye, M. (1990). *Crit. Rev. Biochem. Mol. Biol.*, **25**, 155.
163. Izant, J. G. and Weintraub, H. (1984). *Cell*, **36**, 1007.
164. Groger, R. K., Morrow, D. M., and Tykocinski, M. L. (1989). *Gene*, **81**, 285.
165. Holt, J. T., Venkat-Gopal, T., Moulton, A. D., and Nienhuis, A. W. (1986). *Proc. Natl. Acad. Sci. USA*, **83**, 4794.
166. Yates, J. L., Warren, N., and Sugden, B. (1985). *Nature*, **313**, 812.
167. Graham, F. L. and ven der Eb, A. J. (1973). *Virology*, **52**, 456.
168. Felgner, P. L., Gadek, T. R., Holm, M., Roman, R., Chan, H. W., Wenz, M., Northrop, J. P., and Danielson, G. M. R. (1987). *Proc. Natl. Acad Sci. USA*, **84**, 7413.
169. Neumann, E., Schaefer-Ridder, M., Wang, Y., and Hofschneider, P. H. (1982). *EMBO J.*, **1**, 841.
170. Muller, H. J. (1932). *Proc. 6th Internat. Cong. Genetics*, **1**, 213.
171. Herskowitz, I. (1987). *Nature*, **329**, 219.
172. Kulesh, D. A., Cecena, G., Darmon, Y. M., Vasseur, M., and Oshima, R. G. (1989). *Mol. Cell. Biol.*, **9**, 1553.
173. Trevor, K. T. (1990). *New Biol.*, **2**, 1004.
174. Loewinger, L. and McKeon, F. (1988). *EMBO J.*, **7**, 2301.
175. Letai, A., Coulombe, P., and Fuchs, E. (1990). *J. Cell Biol.*, **111**, 176 (abstract).
176. Hogan, B., Constantini, F., and Lacy, E. (1986). *Manipulating the mouse embryo: a laboratory manual.* Cold Spring Harbor Press, New York.
177. Vassar, R., Coulombe, P. A., Degenstein, L., Albers, K., and Fuchs, E. (1991). *Cell*, **64**, 365.
178. Capetanaki, Y., Smith, S., and Starnes, S. (1989). *J. Cell Biol.*, **109**, 1653.
179. Dunia, I., Pieper, F., Manenti, S., van de Demp, A., Devilliers, G., Benedetti, E. L., and Bloemendal, H. (1990). *Eur. J. Cell Bio.*, **53**, 59.
180. Powell, B. C. and Rogers, G. E. (1990). *EMBO J.*, **9**, 1485.
181. Baribault, H. and Kemler, R. (1989). *Mol. Biol. Med.*, **6**, 481.
182. Baribault, H., Price, J., and Oshima, R. G. (1990). *J. Cell Biol.*, **111**, 40 (abstract).
183. Klymkowsky, M. W., Maynell, L. A., and Nislow, C. (1991). *J. Cell Biol.*, **114**, 787.

Appendix: List of suppliers

(All addresses are in the USA.)

Aldrich Chemical Co., P.O. Box 355, Milwaukee, WI 53201
Amersham Corp., 2636 South Clearbrook, Arlington Heights, IL 60005
Amicon Division, W.R. Grace & Co., 24 Cherry Hill Dr., Danvers, MA 01923
Beckman Instruments, 1050 Page Mill Road, Palo Alto, CA 94304
BioRad Laboratories, 1414 Harbor Way South, Richmond, CA 94804
Biospec Products, P.O. Box 722, Bartlesville, OK 74005
Boehringer-Mannheim Corp., Biochemical Division, 9115 Hague Road, P.O. Box 50414, Indianapolis, IN 46250–0414
R. P. Cargille Laboratories, Inc., Cedar Grove, NJ 07009
Gelman Sciences Inc., 600 S. Wagner Rd., Ann Arbor, MI 48106–1448
DAGE-MTI, Inc., 701 N. Roeske Ave., Michigan City, IN 46360
David Kopf Instruments, P.O. Box 636, 7324 Elmo Street, Tujunga, CA 91042
Delaware Diamond Knives, Inc., 3825 Lancaster Pike, Wilmington, DE 19805
Division of Marine Resources, Marine Biological Laboratories, Woods Hole, MA 02543
Electron Microcopy Sciences, Box 251, Fort Washington, PA 19034
Flow Labs, Inc., 7655 Old Springhouse Road, McLean, VA 22102
Frederick Haer & Co., Brunswick, ME 04011
Gulf Specimens, Box 237, Panacea, FL 32346
Hamamatsu Photonic Systems Corp., 2625 Butterfield Road, Oak Brook, IL 60521
ICN Biochemicals, Inc., Research Products Div., P.O. Box 5023, Costa Mesa, CA 92626
Inovision Corp., P.O. Box 12539, 2810 Meridian Parkway, Suite 100, Research Triangle Park, NC 27709
Ken-Tex Machine Shop, Route 1, P.O. Box 146, Trenton, KY 42286
Life Cell Corp., 3606–A Research Forest Drive, The Woodlands, TX 77381
Molecular Probes, Inc., P.O. Box 22010, 4949 Pitchford Avenue, Eugene, OR 97402
Newport Corp., P.O. Box 8020, 18235 Mt. Baldy Circle, Fountain Valley, CA 92728
New England Miniature Ball Company, Norfolk, CT
Numonics Corp. 101 Commerce Dr., Montgomery, PA 18926
Organon Teknika-Cappel, 100 Akso Ave., Durham, NC 27704

Pharmacia LKB Biotechnology Inc., 800 Centennial Ave., P.O. Box 1327, Piscataway, NJ 08855–1327
Photometrics Ltd., 3440 E. Britannia Dr., Tucson, AZ 85706
Pierce Chemical Co., P.O. Box 117, Rockford, IL 61105
Promega Biotech., 2800 Fish Hatchery Road, Madison, WI 53711
Schleicher and Schuell, Inc., 10 Optical Dr., Keene, NH 03431
Sigma Chemical Co., P.O. Box 14508, St. Louis, MO 63178
TAGO, Inc., 887 Mitten Rd., P.O. Box 4463, Burlingame, CA 94010
Technical Video, Ltd., P.O. Box 693, Woods Hole, MA 02543
Tekmatic Systems, 550 Hillcrest Drive, Ben Lomond, CA 95005
Whatman Inc., 9 Bridewell Place, Clifton, NJ 07014
TSI Inc., 500 Cardigan Rd., P.O. Box 64394, St. Paul, MN 55164
Universal Imaging, 502 Brandywine Parkway, Westchester, PA 19380
Upstate Biotechnology, Inc., 89 Saranac Avenue, Lake Placid, NY 12946
VWR Scientific, P.O. Box 13645, Philadelphia, PA 19101–3645
YSI Scientific, Yellow Springs, OH 45387
Carl Zeiss, Inc., One Zeiss Dr., Thornwood, NY 10594

Index

Index

in microtubule dynamic analysis 153–4, 172
in Nomarski microscopy analysis 153–4, 172
concanavalin A, membrane glycoprotein
    studies employing 132, 133
conformation/folding (of proteins)
    in bacterial expression systems 105–6
    calcium induced changes in 79–80
contrast enhancement of fluorescence images
    17
cover slips
    microtubule motility assays regarding
        quality of 193
    wet cleavage of cells between two 27–8
critical-point drying 34
CSAT (β-integrin; fibronectin receptor)
    actinin and, interactions 139
    fibulin and, interactions 139
    talin and, interactions 142
C-terminus of protein, domain mapping with
    antisera recognizing 80–3
cultures, cell
    bacterial
        for actin expression 112
        for intermediate filament protein
            expression 228–9
    for fluorescent analogue cytochemistry
        8–12
    yeast 200–1
cyanogen bromide cleavage sites, domain
    mapping at 85, 86–7
cysteine residues
    in actin-binding proteins, domain mapping
        with cleavage at 84–5
    of intermediate filament proteins 225, 237
cytochalasin D, actin polymerization inhibited
    by 69
cytochemistry, fluorescent analogue, *see*
    fluorescent analogues
cytokeratins 239
    antimorphic mutations 246–7, 247, 248
    assembly 234–5
cytoplasm
    intermediate filament 223
    peripheral protein 127
cytoskeleton, definition/constituents 23, 125

denaturing reagents in solubilization of inter-
    mediate filaments 226–7
deoxyUTP incorporation in site-directed
    mutagenesis 102–3
depolymerization
    of actin filaments 53–6
    of intermediate filaments 230
detergents, membranes solubilization/
    permeabilization with 24–8, 128,
    215–16
    in yeast 215–16

Dictostelium discoideum actin expression in
    bacterial systems 105–6, 116
diethylene glycol distearate, sections
    embedded in 38
differential-interference-contrast microscopy
    in microtubule dynamic studies
    151–66, 170–5
distillation, molecular 35
DNA
    extrachromosomal, mutants introduced via
        202–4, 205
    genomic
        hybridization analysis 213–14
        isolation 212–13
    cDNA, actin cloned in 100
    ssDNA (single-stranded DNA), intermediate
        filament protein chromatography
        employing 231–2
DNase I, inhibiition by G-actin 67–8
DNase I-affinity chromatography of mutant
    actins 116–17, 119–20, 120
    separation from wild-type actins 120
DNase I-agarose, preparation 116–17
domains
    of actin-binding proteins, *see* actin-binding
        proteins
    of intermediate filaments, head and tail,
        polymerization and 234
dried specimens, metal coating 31–3
Drosophila
    flight muscle, mutant actin expression in
        106, 109
    ncd gene coding for kinesin-like protein 194
drug-resistant yeast mutants, selection 201–2
dry cleavage of cytoskeleton 34–5
drying
    critical-point 34
    freeze- 31–4
dynamics
    cytoskeletal (in general), fluorescence
        microscopic analysis of 1–21
    intermediate filament 237–8, 240–2
    microtubule 65, 151–96
    see also motion
dynein 171, 176, 178–9
    bidirectional movement in axons of
        membrane-bound organelles and role
        of 171
    isolation 178–9
    with kinesin 179, 180
    microtubule-based motility induced by
        179–86 *passim*

electron microscopy, *see* ultrastructural studies
electrophoresis, SDS polyacrylamide gel, *see*
    SDS-PAGE
embedding of cells 37–8

261

Index

Index